P9-DZZ-056

OBSERVATION AND THEORY IN SCIENCE

THE ALVIN AND FANNY BLAUSTEIN
THALHEIMER LECTURES

1969

MAURICE MANDELBAUM
Editor

Observation and Theory in Science

ERNEST NAGEL
SYLVAIN BROMBERGER
ADOLF GRÜNBAUM

With an Introduction by
Stephen F. Barker

THE JOHNS HOPKINS PRESS BALTIMORE AND LONDON

Manufactured in the United States of America

The Johns Hopkins Press, Baltimore, Maryland 21218
The Johns Hopkins Press Ltd., London

Library of Congress Catalog Card Number 72-150043
International Standard Book Number 0-8018-1303-4

Contents

OBSERVATION AND THEORY IN SCIENCE

Introduction

The three lectures of which this volume consists were delivered in Baltimore in April and May of 1969, under the auspices of the Department of Philosophy at The Johns Hopkins University, as the first series of the Alvin and Fanny Blaustein Thalheimer Lectures. This lecture series has been made possible by a gift from the Thalheimer Foundation, for which the Department is extremely grateful. That gift will permit, for some time to come, the presentation of further series of lectures on a variety of philosophical topics.

The three lectures of this first series deal with different aspects of a single theme—the logical status of scientific theories in their relation to observation. The lectures treat this theme in terms of contemporary discussion and in connection with some recent controversies in the philosophy of science. A non-specialist who encounters these lectures should realize, however, that the theme itself is a perennial one, with an ancient lineage. It has concerned philosophers from the earliest era of philosophy on down through the centuries.

A central philosophical issue that is at stake here

might be expressed in this way: Are scientific theories testable in terms of our observations in such a manner that we can know some theories to be true and some theories to be false? Or is the logical connection between theories and observations so remote and so flexible that theories are always immune to any definite proof or refutation through observation? The answer that definite testing of theories by means of observations can occur is an answer which harmonizes with the realistic view of science, the view that those scientific theories which best conform to observed facts are true and afford us genuine knowledge of how things are. This is the view most congenial to common sense and to the usual outlook of scientific philosophy. On the other hand, the answer that theories are immune to such testing would tend to support a sceptical or idealistic outlook, according to which scientific theories are incapable of being true or false in any literal sense and cannot afford us knowledge of the world as it is—though perhaps they serve some other useful purpose, such as to facilitate predictions concerning observable phenomena.

The issue here is not just a matter of general confidence in human knowledge *versus* general scepticism about man's ability to know anything. Instead, it arises out of a special factor—the contrast which there seems to be between observing and theorizing. The typical form of scepticism about scientific theories has (at least in the past) based itself upon acceptance of this contrast and upon acceptance of observing as a comparatively secure way of acquiring knowledge; its attack has been directed against theorizing. The main point of this scepticism has been its insistence that scientific theories are not really testable. Scientific theories, unlike observations, seem to involve claims going far beyond what is warranted by sense-experience; these claims are highly speculative in character, essentially incapable of being established, and perhaps also incapable of being refuted.

An important further thought, often associated with this scepticism, is its insistence that the very meaningfulness of scientific theories is suspect: for the special terms occurring in scientific theories do not seem to draw their meanings from sense experience. And if theories lack meaningfulness, of course they cannot be known to be true or known to be false. Consider, for instance, a geneticist who observes the colors of successive generations of sweet peas, and who then claims that this supports a set of principles about genes; or a physicist who observes certain tracks in photographs of bubble-chambers, and then claims that these support various propositions about sub-atomic particles. In such cases, the theoretical claims seem to go very far beyond what observation directly establishes, thus inviting scepticism about whether we can ever know that these claims are true or false. Moreover, it is puzzling how terms like "gene" and "electron" can be meaningful to us, since these are notions of unobserved entities which we never confront directly in our experience.

The sceptical interpretation of scientific theories had an early and powerful exponent in Plato. Of course Plato was not a sceptic in general, for he believed that there is genuine knowledge of reality which the human mind can attain; but he held that this is rational knowledge of the eternal Forms, not empirical knowledge of the spatio-temporal world. Plato did not say that the terms in scientific theories lack meaning, but by insisting upon a view of their meaning incompatible with empiricism, he sowed the seeds of later doubt. Plato's view was that the meanings of terms used in stating scientific theories about the physical world are not concepts derived from or genuinely exemplified in sense-experience, but instead are purely rational notions attainable only through the mind's intellectual communion with eternal Forms. The geometrical notions of the regular solids, for instance, are used by Plato in a sort of atomic theory; but he believed that these, like all the distinctive concepts of geometry, are notions, genuine instances of which are never encountered in sense experience, and that the mind can have acquired these notions only from a transcendent source. In the *Timaeus,* where Plato theorizes extensively about the physical world, he seems to hold that the proper task of scientific theories is merely to 'tell a plausible story' and to 'save the appearances,' i.e., to invent theories which are consistent with the observed facts, but which cannot be shown to be true—since many other contrary theories are equally consistent with the observed facts also. Under such circumstances, one's theories of atomic physics or of astronomy cannot be more than sheer groundless conjectures, and no genuine knowledge will be humanly attainable concerning such subject matter.[1]

This, at any rate, is how Plato's position was understood by many of his followers. Such an outlook finds a striking expression in the preface that Osiander wrote

[1] It should be noted, however, that this interpretation of Plato is controversial. For a contrary interpretation of Plato's position on this point, see Karl Popper's discussion of sceptical views of science in his *Conjectures and Refutations* (London: Routledge & Kegan Paul, 1963), p. 99.

for Copernicus's *De Revolutionibus.* Wishing to mollify the antagonism which he feared Copernicus's theory would excite, Osiander wrote that the theory should not be interpreted as saying that the earth really moves; all the theory says is that observations proceed as if the earth were moving, and that this way of regarding matters facilitates calculation of predicted future observations. And Osiander seems to have said this not because he believed that all motion is relative, but rather because he was willing to advocate the view that all scientific theorizing concerning the physical world is unverifiable speculation.

A more recent noteworthy representative of this sceptical outlook concerning theories was the French physicist and historian of science, Pierre Duhem. Duhem is remembered in philosophy today for his thesis that there are no crucial experiments in science. Duhem argued that, in any attempt to bring a scientific theory to observational test, many subsidiary empirical assumptions always have to be made, so that seemingly negative observations always leave open the option of regarding one or another of those subsidiary assumptions rather than the theory itself as false: thus a theory can never be strictly refuted by any experimental observation.[2] However, in Duhem's own philosophy of science, this thesis about the impossibility of crucial experiments is only one element of a larger picture, a stepping stone toward a final scepticism rather like

[2] This view of the relation between theory and observation is summed up in Quine's well-known saying that theories "face the tribunal of sense experience not individually but only as a corporate body." W. V. Quine, *From a Logical Point of View* (Cambridge, Mass.: Harvard University Press, 1953), p. 41.

Plato's concerning the over-all import of scientific theories. Whatever our observations may be, always many conflicting theories are compatible with them; and Duhem says, "If two different theories represent the same facts with the same degree of approximation, physical method considers them as having absolutely the same validity; it does not have the right to dictate our choice."[3] He goes on to add that the physicist will often choose between these competing theories, but that the motives which will dictate his choice will be considerations of elegance, simplicity, and convenience—considerations which Duhem reckons to be essentially subjective. His conclusion is that scientific theories are neither true nor false. They can have no bearing upon objective reality; they cannot affirm or deny that any real being does or does not possess a certain attribute.[4] What theories do is to "summarize and classify laws established by experiment"; theories consist of mathematical formulas "set up by a free decree of our understanding" in order to permit us to deduce a series of consequences furnishing us a more or less satisfactory representation of the laws noted in our laboratories.[5] Here, as with Plato and Osiander, we have a view which seeks to curb the pretensions of scientific knowledge, thereby making room for metaphysical and religious belief—though Duhem is led to his scepticism by consideration of the unfalsifiability of theories, whereas earlier scepticism merely stressed their unverifiability.

In more recent twentieth-century philosophy of

[3] Pierre Duhem, *The Aim and Structure of Physical Theory* (New York: Atheneum, 1962), p. 288.

[4] *Ibid.*, pp. 283, 287.

[5] *Ibid.*, p. 285.

science, the question of the meaningfulness of theoretical terms (which Duhem scarcely touches) was given especial prominence, particularly among philosophers who adopted the viewpoint of 'logical empiricism.' A quite special view of scientific theories as 'partially interpreted formal systems' was developed a generation ago by Carnap[6] and others, and came to be quite widely held. (In his lecture, Bromberger calls it simply "the accepted view," speaking with conscious exaggeration.) The basic philosophical motivation for this view derived from an empiricistic concern about the meaningfulness of the terms which occur in scientific theories. Logical empiricists were inclined to think that all meaningful words and symbols (except for the words and symbols distinctive of logic) must derive their significance from sense experience; that is, must have meanings that are ultimately explainable ostensively, by reference to things that can readily be observed. What they called 'observation terms'—words like "blue," "hard," and "hot"—seemed to meet this requirement fully. But other terms that are important in science—terms like "gene" and "electron" which they called 'theoretical terms'—did not seem to meet it. How then can scientific theories be legitimate? The notion of formal systems was invoked in an attempt to explain this.

The notion of a formal system was borrowed from mathematical logic, where formalism as a method for studying the foundations of mathematics had been achieving notable successes. There, instead of regarding mathematical postulates and theorems as meaningful

[6] See, for example, Rudolf Carnap, *Foundations of Logic and Mathematics* (Chicago: University of Chicago Press, 1939), pp. 56–69.

sentences expressing truths, the approach was to deal with them as mere strings of marks, paying no attention to whether any significance had been assigned to these marks. The rules for combination and recombination of these marks were investigated in order to establish conclusions as to which strings of marks were derivable under the given rules from which others. This approach to mathematics made it possible for investigators to study important logical properties of mathematical systems without being obliged to take any stand concerning whether the mathematical formulas under study were true, or even whether they were meaningful.

This seemed to offer a key for solution of the philosophical problem concerning scientific theories. Let scientific theories be regarded as formal systems also: then their role in science can be discussed without the assumption that the sentences occurring in them are fully meaningful. Scientific theories could be viewed simply as systems whose rules permit the deriving of certain new combinations of words or symbols from given ones. Let the vocabulary of science be divided into observational terms (possessing direct meanings derived from experience) and theoretical terms (possessing no such direct meanings); sentences containing only observational terms (and called 'observation sentences') will be capable of being decisively verified or refuted by sense experience. The role of theories will be to provide logical devices whereby new observational sentences can be derived from ones already at hand—thus yielding suggestions for further observations and predictions concerning their outcome. Theoretical terms, though having no direct meaning, were described as acquiring 'indirect meaning' or 'partial interpretation' through their indirect linkage with observations, via the rules of the system.

Unfortunately, however, this way of trying to reach a philosophical understanding of the status of scientific theories raises more difficulties than it settles. One type of difficulty for this view of theories arises out of the account it tries to offer of why theories are of value to science. Different advocates of this view of theories differ in the extent to which they are willing to embrace the sceptical view that scientific theories are incapable of truth or falsity and cannot embody knowledge of how things are.[7] However, all who advocate this view of theories have wished to defend the activity of theorizing and have wanted to maintain that it has some kind of value. Yet the problem arises: Why is it of any value to introduce theoretical terms into science, if they lack direct meaning and if theories containing them are merely formal systems? What useful purpose can be served by this?

This query became especially pressing in light of the proof by Craig that any derivation of new observational sentences from given ones that can be accomplished in a formal system containing theoretical terms can also be accomplished in a system containing observational terms only and no theoretical terms at all.[8] What contribution then can the theoretical terms make? Some philosophers of science have said vehemently that

[7] Scheffler has classified the different types of opinion on this matter. Israel Scheffler, *The Anatomy of Inquiry* (New York: Alfred A. Knopf, 1963).

[8] William Craig, "Replacement of Auxiliary Expressions," *Philosophical Review* 65 (1956):38–55.

Craig's theorem is of no relevance to the philosophy of science; in a sense, that may be correct, yet the theorem is of great importance as an objection to the view of theories as formal systems. Hempel called this problem "the theoretician's dilemma,"[9] and sought to defend this view of theories by holding that, although from a purely deductive viewpoint theoretical terms cannot contribute anything of value, once one adopts the viewpoint of inductive logic, theoretical terms can be seen to be of use under this conception of theories. His examples of the inductive contribution of theoretical terms were not conclusive, however, and there remains some difficulty here for this point of view.

Another type of difficulty for this view of theories concerns the assumption upon which the whole approach is based, the assumption that there is a definite distinction to be drawn between so-called 'observational terms' whose meanings are abstracted directly from experience and 'theoretical terms' whose meanings, if any, are much more problematic. Carnap, to be sure, was quite willing to admit that it is an arbitrary matter which terms one classifies as 'observational';[10] yet his whole philosophical account of scientific theories, if it is to be regarded as having any philosophically illuminating value, must be viewed as resting on the assumption that the observational–theoretical dichotomy is by no means essentially arbitrary. For if the dichotomy between observational terms and theoretical terms were

[9] C. G. Hempel, "The Theoretician's Dilemma," *Minnesota Studies in the Philosophy of Science II* (Minneapolis: University of Minnesota Press, 1958).

[10] Rudolf Carnap, *Philosophical Foundations of Physics* (New York: Basic Books, Inc., 1966), p. 226.

largely arbitrary, then it could not be the case that the meanings of theoretical terms are somehow essentially different in kind, more suspect, and more in need of a special philosophical explanation than are the meanings of observational terms. And in that event, there would be no more discernible reason for proposing to regard theoretical sentences as partially interpreted than there would be for regarding observational sentences (or any other set of sentences) in that light. The notion of 'partial interpretation,' if it has any value at all, attains its value on the basis of a contrast between terms that have some genuinely straightforward type of meaning and terms whose meaning is genuinely less so. The account of scientific theories as partially interpreted formal systems would be uncalled for and without philosophical point, unless it could be based on the assumption that there were a non-arbitrary distinction between observational terms and theoretical terms.

Critics have attacked this assumption in various ways.[11] The mildest criticism is that the advocates of the formalistic view of scientific theories have failed to explain the distinction adequately, leaving it too vague and ambiguous. More serious is the objection that there

[11] For some of these criticisms, see J. L. Austin, *Sense and Sensibilia* (Oxford: Oxford University Press, 1962); Paul Feyerabend, "Problems of Empiricism," in *Beyond the Edge of Certainty*, University of Pittsburgh Series in the Philosophy of Science, vol. II, ed. Robert G. Colodny (Pittsburgh: University of Pittsburgh Press, 1965); Peter Achinstein, *Concepts of Science* (Baltimore: Johns Hopkins University Press, 1968); and Frederick Dretske, *Seeing* (London: Routledge and Kegan Paul, 1970). An earlier formulation of some criticisms is found in my *Induction and Hypothesis* (Ithaca: Cornell University Press, 1957).

does not exist any such distinction to be drawn—that the whole idea of some terms having meanings 'directly given' in experience is based on misunderstanding, and that all terms employed in science, both for formulating observations and for formulating theories, are alike in sharing essentially the same sort of meaningfulness. Finally, the most extreme criticism of all has arisen in recent years from philosophers such as Paul Feyerabend who argue that the meanings of allegedly 'observational terms' actually are 'theory-laden' in such a fashion that their meanings are affected by and dependent upon the theories to which they pertain. This outlook implies that, far from the theory needing to have its meaning explained in terms of reports of observations, it is rather the observational reports whose meanings require explanation in terms of the theory. Such an outlook, however, suggests that each theory is immune to refutation in its own terms, yet cannot be tested in any but its own terms. And this leads back toward scepticism concerning the possibility of our ever knowing whether a theory is true or whether it is false.

These, at any rate, are some of the strands that have run through discussion in the philosophy of science of the relations between observation and scientific theory. The three Thalheimer lectures of this series contribute to this ongoing discussion by treating specific aspects of the problem.

Professor Nagel in his lecture aims to defend scientific theorizing against sceptical attacks. He analyzes some of the particular attacks that have been made upon the observational-theoretical distinction. Then, going beyond his earlier powerful treatment of these issues,[12] he proposes that the proper way of construing the observational-theoretical contrast is in terms of the way statements actually function within the context of scientific inquiry. Developing this idea will inevitably lead us some distance from the formalistic view of scientific theories as it was conceived by its original proponents. However, Professor Nagel argues that we shall thereby best bring out the real point of their endeavors.

In the second lecture, Professor Bromberger proposes a very different approach. Feeling the formalistic view of theories to be essentially bankrupt, he sketches a radically different way of undertaking the analysis of scientific theories—in terms of questions raised, and in terms of the logical dimensions of possible answers to these questions. Professor Bromberger's account of this alternative approach is only schematic, yet it suggests exciting possibilities for giving us a clearer perspective upon the status of theories, while freeing us from the need to invoke the formalistic apparatus with its attendant difficulties.

In the final lecture, Professor Grünbaum renews a discussion which he had earlier undertaken concerning Duhem's thesis that it is impossible to test scientific theories by crucial experiments. He makes some new distinctions and modifies his judgment concerning the sense in which Duhem's thesis is correct and the sense in which it is incorrect. In connection with this, he seeks to show through his detailed examples how recent philosophers, such as Feyerabend, have misinterpreted the way in which alterations in theoretical principles are re-

[12] Ernest Nagel, *The Structure of Science* (New York: Harcourt, Brace & World, 1961).

lated to changes in the meanings of scientific terms. Professor Grünbaum (agreeing with Nagel) seeks specifically to present counterexamples to Feyerabend's claim concerning the manner in which terms are 'theory-laden.' In this way, Grünbaum, too, is opposing the scepticism inherent in Feyerabend's position.

Thus, although they concentrate on somewhat different issues, all three of these lecturers are dealing with aspects of a single theme. All three are seeking a more plausible and non-sceptical philosophical account of the status of scientific theories in relation to observation.

Stephen F. Barker

Ernest Nagel

Theory and Observation

In his well-known Herbert Spencer Lecture "On the Method of Theoretical Physics," delivered in 1933, Einstein took as the starting point of his discussion what he characterized as "the eternal antithesis" between the rational and the empirical components of physical knowledge, an antithesis he reaffirmed in his subsequent reflections on scientific method. He pointed out that a physical theory is an ordered system of certain fundamental general postulates or assumed laws, stated in terms of various basic concepts dealing with the matters under study, and he stressed the indispensable role of logical deduction in making explicit the numerous detailed consequences of those postulates. He also underscored his conviction that while such concepts and assumptions may be, and perhaps usually are, *suggested* by the things investigated in physics, they are not derivable from those things either by abstraction or by some other logical process, but are "free inventions of the human mind." However, Einstein went on to say, the imaginative creation of ideas and "purely logical thinking cannot yield us any knowledge of the empirical world; all knowledge of reality starts from experience and ends with it. Propositions arrived at by purely logical means are completely empty as regards reality." If a physical theory is to count as a valid account of the actual world, and to be more than the formulation of logical possibilities, the conclusions deduced from its postulates "must correspond" with what experience reveals. "Because Galileo saw this," Einstein continued, "and particularly because he drummed it into the scientific world, he is the father of modern physics—indeed, of modern science altogether."[1] Accordingly, although on Einstein's view neither the basic concepts nor the fundamental postulates of a theory are uniquely prescribed or determined by observation and experiment, and cannot therefore be simply "read off" from what is encountered in sensory experience, the *validity* of a theory as an account of the actual world depends upon, and is controlled by, what is disclosed to observation and experiment.

In emphasizing the decisive role of these latter in the *testing* of physical theories, Einstein was stating a view that has been pretty much of a commonplace for several centuries in discussions of scientific method. Even his claim that the central ideas of modern physical theory are neither abstracted from, nor explicitly definable with finitistic means in terms of, unquestionably manifest traits of the phenomena studied in the science, is at present not a seriously mooted issue among philosophers—though undoubtedly the claim contradicts beliefs associated with certain historical forms of empirical philosophies of knowledge and continues to be included

[1] Albert Einstein, *Ideas and Opinions* (New York: Crown Publishers, 1954), pp. 271–72.

in widespread popular notions about the nature of scientific inquiry. In any event, there are no current analyses of the logic of inquiry in the natural sciences which formally deny the cardinal role of what Einstein called "experience" (i.e., controlled observation and experiment) in assessing the cognitive worth of a scientific theory.

However, despite the absence of such formal denial, the authority of experience in evaluating the adequacy of theoretical assumptions has nonetheless been recently challenged. For the view that a proposed theory in the natural sciences must be tested against the findings of experience appears to rest on the assumption that a clear distinction can be drawn between, on the one hand, statements which supposedly codify the outcome of observations on the subject matters being explored and, on the other hand, the theories about those matters for which the so-called "observation statements" supposedly provide confirming or disconfirming evidence. Moreover, many philosophers of science who think that such a distinction can be instituted with at least approximate precision also believe that observation and theoretical statements frequently have no non-logical expressions (or subject matter terms) in common; and they therefore maintain that, if such purely theoretical statements are to be relevant to empirical subject matters, theoretical statements must be related to observation ones by way of so-called "correspondence" rules or laws. But these assumptions, especially the first one, have been the targets of vigorous criticism during the past decade, criticisms which have in fact reopened the question whether, and if so to what extent, theories are subject to the test of experience.

The main burden of these criticisms is that the belief in an "absolute" distinction between observation and theory is untenable. Observation statements, so the criticisms commonly maintain, are not unbiased formulations of supposedly "pure" materials of sensory experience, but involve interpretations placed upon the sensory data, and are therefore significant only in virtue of some theory about the things under study to which the observers are antecedently (though not necessarily permanently) committed. Moreover, theories are ostensibly "free creations" of the scientist, and while their adoption may have definite causal determinants, none of them is logically prescribed by the sensory data. The import of every observation statement is therefore determined by some theory that is accepted by the investigator, so that the adequacy of a theory cannot be judged in the light of theory-neutral observation statements. Accordingly, if these criticisms are sound, they apparently lead to a far-reaching "relativism of knowledge," to a scepticism concerning the possibility of achieving warranted knowledge of nature that is much more radical than the relativism associated with the views of Karl Mannheim and other sociologists of knowledge. For according to Mannheim, for example, the content as well as the standards for evaluating the validity of a theory in the social sciences (though not in the natural sciences) reflect the social biases and vary with the social perspectives of the individuals who subscribe to the theory. These are, nevertheless, biases which are generated by causal factors operating in certain branches of inquiry, biases which even Mannheim recognized could in principle be neutralized. But the relativity of knowledge implicit in some recent trends in the philosophy of science is of a more fundamental sort. The scepticism it entails concerning the possibility of deciding on objective observational grounds between alternative theories, is based on its analysis of the intrinsic structure of human cognitive processes.

But however this may be, this paper is a critique of some of the reasons that have been recently advanced for questioning the authority of even painstaking observation in testing theories, and for supporting a view of science that makes genuine knowledge of nature highly problematic. This sceptical import of recent discussions in the philosophy of science seems antecedently incredible, and indeed I hope to show that the evidence for such a conclusion is far from compelling. It would be idle to pretend, however, that there are no difficulties in drawing a distinction between observational and theoretical statements; and I certainly do not know how to make such a distinction precise. Nevertheless, I do not consider that this distinction is therefore otiose any more than I believe that the fact that no sharp line can be drawn to mark off day from night or living organisms from inanimate systems makes these distinctions empty and useless.

At the same time, I readily admit that the position for which I shall be arguing is essentially a middle-of-the-road one, and is certainly not novel. For on the one hand, I agree with the view repeatedly advanced in the history of thought by critics of sensationalistic empiricism—though proclaimed as a novel idea by some recent commentators on the logic of science—that the sense and use of predicates employed in the sciences, including those employed to report allegedly observed matters, is determined by the general laws and rules into which those predicates enter. In consequence, the content of an observation statement is not identifiable with or exhausted by what is "directly" encountered in any given sensory experience, so that every such statement is corrigible "in principle" and may be revised (and perhaps even totally rejected) in the light of further observation and reflection. On the other hand, I also think that many observation statements acknowledged in controlled inquiry as well as in the normal affairs of life do not *in fact* require to be corrected; that the various laws which determine in whole or in part the content of such statements are often so well supported by evidence that they are beyond reasonable doubt; and that the content of observation statements is not in actual fact determined by the *totality* of laws and rules of application belonging to the corpus of assumptions of a science at a given time. But if this is so, the meaning and validity of observation statements are in general not determined by the theory that those statements are intended to test; and accordingly, such statements can be used without a stultifying circularity to assess the factual adequacy of the theory. However, these claims need to be supported by argument, a task to which I now turn.

1

The words "theory" and "observation" as well as their various derivatives are notoriously ambiguous as well as vague. Indeed, although the expressions "theoretical statement" and "observational statement" are often associated with ostensibly contrasting senses, there are also contexts in which they are used with overlapping denotations. For example, the adjective "theoretical" is sometimes employed to refer to statements about microentities or microprocesses, such as statements about the motions of electrons, on the assumption that these matters cannot be perceived (in the sense in which an experimental psychologist understands the word "perception"), while the term "observational" is reserved for reports of macro-occurrences that can be so perceived, such as the sounds emitted by some laboratory instrument. On the other hand, statements concerning the distribution of electrons on an insulated conductor or concerning the surface temperature of the sun are sometimes designated as observational, despite the fact that the matters reported are not perceived in the sense indicated. Moreover, the word "theoretical" is frequently used more or less interchangeably with "conjectural," as when a tentative hypothesis as to how a robbery was committed is said to be a theoretical account of the event; and in this case the possibility is not excluded that the conjectured occurrences are also observable in a relatively strict sense of the word, so that they might have been recorded in a suitable set of observation statements. Again, some writers understand by "theory" any essentially general statement, such as "All blue-eyed human parents have blue-eyed children," even if the predicates contained in it designate matters that are usually counted as observable ones. Other writers reserve the word for a system of general assumptions capable of explaining (and perhaps also predicting) the occurrence of a wide range of diverse phenomena, such as the assumptions that constitute Newtonian gravitational theory; and there are some thinkers who use the word rather loosely for the totality of general beliefs, attitudes, or categorial distinctions that make up a *Welt-*

anschauung (or, in current idiom, a "conceptual framework" and even a "form of life"). The word "observable" has a comparable diversity of usages, as the above comments suggest.

In any case, it is difficult to formulate with precision the various ways in which the terms "theory" and "observation" are employed. But the senses of these terms that are relevant to the logic of science, and in particular to the central question discussed in this paper, can be illustrated readily enough by typical examples of inquiry in the theoretical sciences. This is not the occasion for an extensive sampling of such specimens, and a single example of what I believe is a representative report of a scientific investigation must suffice. My example is Newton's account of some of the optical experiments he performed in 1666 and of the theory he proposed to explain what he claimed to have observed.

Newton's first letter to the Royal Society containing his "New Theory about Light and Colours" begins with a description of one of his important experiments:

> . . . I procured me a Triangular glass-Prisme, to try therewith the celebrated *Phenomena of Colours*. And in order thereto having darkened my chamber, and made a small hole in my window-shuts, to let in a convenient quantity of the Suns light, I placed my Prisme at its entrance, that it might be thereby refracted to the opposite wall. It was at first a very pleasing divertisement, to view the vivid and intense colours produced thereby; but after a while applying myself to consider them more circumspectly, I became surprised to see them in an *oblong* form; which, according to the received laws of Refraction, I expected should have been *circular*.

> They were terminated at the sides with straight lines, but at the ends, the decay of light was so gradual, that it was difficult to determine justly, what was their figure; yet they seemed *semicircular.*[2]

Newton then went on to report a number of measurements and further observations he made to assure himself that various antecedently plausible sources of the oblong shape of the colored spectrum were not in fact operative. In consequence, he concluded:

> Light is not similar, or homogeneal, but consists of *difform* Rays, some of which are more refrangible than others: So that of those, which are alike incident on the same medium, some shall be more refracted than others, and that not by any virtue of the glass, or other external cause, but from a predisposition, which every particular Ray hath to suffer a particular degree of Refraction.

And he further maintained that

> As the Rays of light differ in degrees of Refrangibility, so they also differ in their disposition to exhibit this or that particular colour. Colours are not *Qualifications of Light,* derived from Refractions, or Reflections of natural Bodies (as 'tis generally believed), but *Original* and *connate properties,* which in diverse Rays are diverse. . . . To the same degree of Refrangibility ever belongs the same colour, and to the same colour ever belongs the same degree of refrangibility. . . . The species of colour, and degree of Refrangibility proper to any

[2] *Philosophical Transactions of the Royal Society,* vol. 6 (1671/72), reprinted in *Isaac Newton's Papers and Letters on Natural Philosophy,* I. B. Cohen, ed., (Cambridge, Mass.: Harvard University Press, 1958), pp. 47–48.

> particular sort of Rays, is not mutable by Refraction, nor by Reflection from natural bodies, nor by any other cause, that I could yet observe. . . .[3]

These quotations will suffice for my purpose. They show beyond doubt that, in describing the observations he made in performing his optical experiments, Newton employed many terms that are "theory laden" in one clear sense of this currently fashionable phrase. Thus, he characterized certain objects as made of glass having the shape of triangular prisms, others as window shutters provided with a small hole, and so on. These predications were not simply reports of sense data "immediately given" to Newton at the time he was carrying on his experiments. On the contrary, the characterizations he used connoted various features of things other than those he explicitly noted as being present—e.g., reflective properties of glass prisms, geometric traits of triangular surfaces, or the opacity of the material out of which the window shutters were constructed. Accordingly, the observation terms used by Newton were not just labels whose contents were exhausted by the directly manifest things to which he applied them, but acquired at least part of their significance from various laws in which they were embedded and which he took for granted. The quotations thus amply confirm what has been familiar to students of human psychology for a long time: that significant observation involves more than noting what is immediately present to the organs of sense; and that a scientific observer comes to his task of reporting his experiments with classificatory

[3] *Ibid.*, p. 53.

schemes representing structures of relations embodied in the flow of events, where the schemes of classification had been acquired in the course of repeated interaction with the environment in ways difficult to recover and enumerate.

However, Newton explained the "phenomena of colours" by a theory that maintained, among other things, that light is composed of distinct "ray," with each ray corresponding to a different color in the spectrum of the sun and having its own inherent degree of refrangibility. But as the above quotations make evident, these theoretical notions so distinctive of Newton's explanation were *not* assumed by him in describing the observations he made when performing his experiments. Moreover, neither his entire letter to the Royal Society nor his much later *Opticks* supplies any ground for holding either that Newton's theoretical explanation of optical phenomena determined (or otherwise entered into) the meanings of the terms he employed in reporting his experimental observations, or that if he had adopted a different explanatory theory those observation terms would have acquired altered meanings.[4] In short, though the predicates which formulated Newton's observations were "theory laden" in the sense indicated, their burden was not the theory he advanced to explain the "phenomena of colours."

[4] There are good reasons for believing that even Newton's theoretical notion of a light ray is independent of various further assumptions that were available to him concerning the nature of light—e.g., the assumption that light is corpuscular, that it is undulatory, or that it travels through an all-pervading optical medium. Cf. Robert Palter, "Newton and the Inductive Method," *The Texas Quarterly* X, no. 3 (1967): 168.

A single example doubtless proves little, and I am not supposing that the passages I have quoted from Newton settle the issue under discussion. But in areas of philosophical analysis in which relatively concrete data are not generally used to test some disputed thesis, even a single bit of such concrete evidence may be useful. In any case, the cited example—and only space forbids mentioning others—does confirm the position I am defending, although it certainly does not prove it. The example shows that an experiment intended to ascertain on what factors the occurrence of a certain phenomenon depends can be described so that the statement of the observations made is neutral as between alternative theories which may be proposed to explain the phenomenon, even though the descriptive statement will indeed presuppose various theories, laws, and other background information that are not in dispute in the given inquiry.

2

Numerous objections have been urged against this *prima facie* plausible thesis. In the main, they are directed against the familiar methodological principle that a theory in empirical science must be tested by confronting it with the findings of observation as codified in so-called "observation statements"; and they are at bottom simply variations on the central contention that the theory-observation distinction is untenable, because even ostensibly "pure" observation statements are in fact impregnated with theoretical notions. However, this fundamental criticism is expressed in recent publications in a number of specialized forms,[5] each emphasizing a different facet of the question, and I will therefore discuss some of the more important versions.

One variant of the criticism maintains that the familiar distinction between theoretical and observation predicates (or between theoretical and observation statements) rests on two tacit and closely related assumptions, neither of which is sound. The first presupposition is that theoretical predicates are inherently opaque and hence problematic, so that if their meanings are to be adequately understood they must be explicated in terms of observation predicates, whose meanings are held to be fully intelligible in their own right. But this assumption is declared to be mistaken. For the meanings of observation predicates, so runs the criticism, far from being luminous and unproblematic, are determined by the numerous statements—and in particular, by the frequently comprehensive theories—into which those predicates enter. The point of this objection is stated by Feyerabend in what is perhaps its most extreme form when he asserts that "theories are meaningful independent of observations. . . . It is the observation

[5] For example, Paul Feyerabend, "Problems of Empiricism," in *Beyond the Edge of Certainty*, Robert G. Colodny, ed. (Englewood Cliffs, N.J.: Prentice-Hall, 1965); N. R. Hanson, *Patterns of Discovery* (Cambridge: Cambridge University Press, 1958); Mary Hesse, "Theory and Observation: Is There an Independent Observation Language?", to be published in vol. 4 of the University of Pittsburgh Series in the Philosophy of Science, Robert G. Colodny, ed.; Thomas S. Kuhn, *The Structure of Scientific Revolutions* (Chicago: University of Chicago Press, 1962); Stephen Toulmin, *Foresight and Understanding* (Bloomington: University of Indiana Press, 1961). In what follows, I have made extensive use of Mary Hesse's essay, and I am indebted to her for formulations of some of the issues.

sentence that is in need of interpretation, and not the theory."[6] The second alleged presupposition underlying the theory–observation distinction is that theoretical science—and especially theoretical natural science—involves the use of two radically disparate *languages*. One of these is a self-contained and autonomous language of observation, whose statements deal exclusively with directly observable matters; the other is a language of theory, whose sentences are explicitly about unobservable things and processes, but are nevertheless significant for scientific inquiry only because of their dependent connections with the observation language. But this assumption is also judged to be unsound. For it is not possible to identify in theoretical inquiry two disparate languages of the sort described, so it is argued, but at best only a *single* language between whose various nonlogical expressions distinctions can be drawn, based on differences in their uses or functions.

There doubtless are philosophers who formulate the distinction between theoretical and observation statements in terms of differences between two supposedly distinct *languages*, so that if taken strictly at their word they are legitimate targets for this criticism. However, not all who subscribe to the distinction state it in those terms; and in any case, the double-edged criticism I have been summarizing seems to me to rest on a misconception of the intent of those making the distinction, even when they employ the locution of "two languages" in formulating it. Thus, there is little if any evidence to show that those accepting the distinction generally maintain that there is an "inherent" difference between

[6] Feyerabend, "Problems of Empiricism," p. 213.

two sorts of predicates or other linguistic expressions, *irrespective* of the uses to which the expressions are put in various situations. On the contrary, it would appear that the rationale controlling the classification of words like "table" and "electron" as observation terms or as theoretical terms respectively, is that these predicates have recognizably different roles in conducting inquiries (and perhaps even in codifying their conclusions). For example, observation terms are commonly used in experimental investigations for purposes such as the following: to mark off in perceptual experience some spatio-temporally located object or process, that may then be subjected to further physical or intellectual analysis; to characterize an item so identified as one of a certain kind; to describe instruments employed in experiments and what is done with them; to state the outcomes of overt measurements and other perceptual findings, so as to provide instantial premises in inferences involving the application of assumed laws to concrete subject-matters; or to codify experimentally ascertained data, with a view of providing tests for general hypotheses and other statements reached by inference in inquiry.

On the other hand, theoretical terms and statements play quite different roles in scientific investigations, often characteristically instrumental ones in mathematical physics. For example, theoretical terms sometimes codify highly idealized (or "limiting") notions, such as the notion of an instantaneous velocity or a point-mass, introduced to simplify intellectual constructions or to make possible the application of powerful tools of calculation to the mathematically "imperfect" materials of the natural world. But without going further into such technical matters, it can be said that theoretical expressions have two major functions in scientific investigations: to *prescribe* how the things identified in gross experience with the help of observation terms are to be analyzed (or otherwise manipulated), if the investigations are to be successful; and to serve as *links* in the inferential chains that connect the instantial experimental data with the generalized as well as the instantial conclusions of inquiry.

This brief account of the functional differences between observation and theoretical expressions is admittedly no more than a sketch. But even this sketch suffices to show that the distinction neither presupposes nor is necessarily committed to the view that observation terms (such as "table" or even "red") are invariably clear, while theoretical ones (such as "electron" or "mass") are inherently problematic. Thus, there are numberless contexts in which neither the meaning nor the applicability of the predicate "table" to some given object raises any issue. But there also are situations in which the applicability of the term is uncertain—for example, when the object in question falls into the area associated with the penumbra of vagueness of the word. Analogously, the term "electron" is unproblematic in many contexts (such as the one in which Millikan performed his oil-drop experiment), when the supposed properties of electrons believed to be relevant to the questions under investigation have been articulated with sufficient precision by some theory of electrons. However, both the meaning and the extension of the term may be quite problematic when the accepted theory of electrons does not determine unequivocally whether electrons possess some stated property—for example, when the question is raised within the framework of current quantum mechanics whether, and if so in what sense, electrons are "particles" rather than "waves." In short, the history of science amply testifies that neither observation expressions nor theoretical ones have invariably unproblematic meanings and denotations, and that expressions of both kinds frequently generate difficult questions of interpretation.

One further remark on the "two languages" locution some writers use in stating the observation–theory distinction. It is pertinent to note that the word "language" in this context does not have the sense ordinarily associated with it when, for example, English or French are called languages. For in this context the word signifies some highly formalized system of notation that is operated in accordance with strict rules. Such systems are undoubtedly valuable for achieving the purposes for which they were devised—for example, to make evident certain structural aspects of everyday discourse, to codify standardized procedures for checking the validity of arguments, or to present in precise form certain distinctions. But such systems are in general not suitable (or even adequate) as total substitutes for natural languages in conducting scientific inquiry as well as the daily affairs of life—for example, for communicating the frequently vague ideas that control investigations, or for describing effectively the often complex but unanalyzed operations involved in research. Accordingly, the "two languages" locution should be construed as a pedagogic device for *distinguishing* analytically between important functions of certain groups of expressions employed in the unformalized discourse of inquiry, rather than as a *descriptive account* of two radically distinct languages between which scientists supposedly shuttle. But if this is the proper construal of the locution, the criticism that the observation–theory distinction assumes that observation terms are uniformly unproblematic is as much a gross exaggeration as is the counter claim that there is never a need to analyze theoretical expressions in order to determine their relations to observation terms.

3

Some writers who subscribe to the distinction have suggested that observation terms can be demarcated from theoretical ones with the help of such alleged facts as that the former but not the latter can be predicated of things on the strength of direct observation, or that the former but not the latter have experimentally identifiable instances. However, critics of the distinction deny this to be the case. They maintain that the meaning and use of a new predicate (whether observational or theoretical) cannot be effectively learned or understood on the basis of "direct experimental association alone," and that these things can be accomplished only by presenting the predicate in various sentences containing other descriptive terms whose meanings had been learned previously. But while this claim may be sound, is it really incompatible with keeping the distinction? As was indicated earlier, what is designated by the observation predicates generally employed in experimental research is not simply the ineffable content of any momentary experience, for they signify characteristics that involve more than what may be immediately given at some time. However, as we also noted earlier, this circumstance is fully consistent with the observation–theory distinction. Accordingly, the criticism under discussion is fatal to the distinction only if observation predicates are equated with so-called "phenomenological" ones—with descriptive predicates that supposedly refer exclusively to what is directly present to the individual who is applying them to matters in his experience. Indeed, if it did presuppose the existence of a purely phenomenological language—that is, one whose descriptive terms are all phenomenological—the distinction would be quite otiose. For no such language is in fact available, and it is a much mooted question whether one is even possible.

Another aspect of the above criticism has also been repeatedly emphasized. This further contention is that the meanings and uses of observation predicates (even of such "basic" ones as "red" and "hard") depend on numerous laws into which the predicates enter, since the laws state, among other things, how the items denoted by the predicates are related to one another. However, the content of these laws is not absolutely invariant, but alters with changes that may occur anywhere in the inclusive network of laws and theories constituting the corpus of scientific knowledge at a given time. In consequence, the meanings of observation predicates are also modified with changes in that network, so that every attempt to distinguish between observation and theoretical terms is bound to fail. Moreover, critics of the distinction reject as unsatisfactory a frequent and plausible rejoinder to this argument. The rejoinder admits that basic terms like "red" may have "peripheral" uses that may be changed because of changes in the network of laws—for example, the apparent redness of a remote star may come to be regarded not as the color of the star but as an effect of its motion. Nevertheless, so the rejoinder continues, it is difficult to make sense of the supposition that changes in the network invariably affect the "core" meanings of all such terms—for example, in the context of predications concerning things like apples or traffic signals, "red" has an apparently stable meaning that is unaffected by changes in most parts of the network of laws. In rebuttal to these contentions, however, one judicious critic of the distinction has argued that while the stability in limited areas of discourse of the core meanings (or functions) of such predicates as "red" must be recognized, the stability "depends not upon fixed stipulations regarding the use of 'red' in empirical situations, but rather upon empirical facts about the way the world is." The conclusion is therefore drawn that physically possible circumstances can be conceived in which even the core meaning of the predicate would no longer be applicable.[7]

This is a puzzling argument. It is doubtless true, as the argument claims, that "the comparative stability of function of the so-called observation predicates is logically speaking an accident of the way the world is." However, the stability may be genuine and important, without being cosmically necessary or unalterable. Accordingly, the contention involved in the observation–theory distinction that observation predicates do have invariant meanings (even if only in restricted domains of application), is surely not refuted by pointing out that things *could* be different from what they are, and that the circumstances on which the invariances depend

[7] Hesse, "Theory and Observation . . . ."

could disappear. Moreover, advances in science undoubtedly affect the way men talk about the world. But it is not inconsistent to admit that scientific knowledge continues to change, and also to maintain that there is a numerous class of predicates whose meanings or functions in various regions of experience (though demarcated only imprecisely) undergo no significant modifications. It is in part because observation predicates do have such relatively constant meanings that the observation–theory distinction is both warranted and useful.

4

A somewhat different reason is sometimes given for challenging the soundness of the distinction. It is pointed out that predicates commonly classified as theoretical, but which belong to the essential vocabulary of a well-entrenched theory, are often used to characterize matters directly apprehended in experimental situations. For example, certain visible tracks in a cloud chamber may be described as the production of a positron–electron pair; a certain land formation may be characterized as a glaciation; and a person who is observed to walk in a specified manner may be described as having a heart ailment. However, if admittedly theoretical predicates can serve as observation ones, the former cannot be distinguished from the latter on the ground that observation terms characterize what is directly observable while theoretical ones do not. Accordingly, there is no absolute difference in this respect between two classes of predicates, but at best only a gradated one.

It is beyond dispute that theoretical terms are sometimes used in the manner stated. The question nevertheless remains whether this fact impairs the validity of the observation–theory distinction. To begin with, it should be noted that many terms often held to be theoretical are apparently never employed to describe ostensibly observable things. Thus, some occurrences are commonly characterized in terms of their causes, but an analogous account of other occurrences would generally be deemed inappropriate. For example, certain sounds may be described as cannon fire, even though the firing of cannons is not being observed; and the click made by a Geiger counter may be described as the passage of an electron, despite the fact that the passing electron is not seen or otherwise perceived. However, in reporting other occurrences nothing analogous is done. For example, a witness in a criminal court would not be permitted to testify that he *observed* someone fire a gun at the deceased, if all he had actually seen was the victim lying on the floor with blood oozing from a hole in his chest. Nor are there seriously intended accounts of what is observed in the electrolysis of water which consist of statements describing rearrangements of electrons in hydrogen and oxygen atoms.

Just why some theoretical expressions are employed to describe observable occurrences (while others apparently never are) is in general unclear as well as controversial. But for many instances of such use of theoretical predicates (as in the case of the term "pair production"), there is a reasonably satisfactory explanation. According to it, the theoretical term in these instances functions as a shorthand formula (based on widely accepted conventions in some branch of inquiry) for characterizing in an effective and discriminating way certain observable but complex features of an experimentally specified occurrence and for distinguishing them from others. For example, the vapor condensations observable in a cloud chamber which, according to accepted physical theory, are effects of pair productions, differ in a number of ways—in shape and thickness, among others—from vapor tracks formed by the passage of alpha particles. However, a long and involved account would be needed if ordinary non-technical locutions were used to characterize adequately the distinctive traits of the observed vapor condensations in these cases. But since the condensations are associated by current physical theory with the occurrence of certain assumed microprocesses, those acquainted with the theory frequently find it more convenient to use the theoretical predicates rather than the more customary terms of perceptual experience in stating what is observed in cloud chambers.

Accordingly, on this explanation, it is only in a Pickwickian sense that theoretical predicates can be counted as observation terms. Even critics of the observation–theory distinction recognize that descriptions of experimental findings couched in theoretical terms (e.g., that in a cloud chamber pair productions occurred) must be replaced by reports employing predicates of normal perceptual experience (e.g., that white tracks were formed), if the theory justifying that use of the theoretical terms is rejected or even seriously challenged. To be sure, the second description, like the first, asserts more than what is directly apprehended by the experimenter, so that the reason for such a replacement clearly cannot be that no error can possibly occur in using the more familiar observation predicates. The reason for the replacement is that the second description asserts things whose existence is better warranted by the actual evidence than are the matters stated in the initial one. On pain of being totally irrelevant to the purposes of performing experiments, theoretical formulations of experimental findings must cover what can be described in ordinary observation terms; but on pain of being wholly superfluous, the theoretical formulations must also assert, if only by implication, things not stated by the latter description. Therefore, when the theory on which such use of theoretical predicates is based becomes doubtful, and when the actual outcome of an experiment is in question and must be established, reports of experimental findings stated in non-controversial observation terms play a crucial role in the conduct of inquiry.

5

The preceding discussion has tried to show that even though observation predicates are in a sense theory impregnated, this thesis is compatible with the observation–theory distinction and is not a good reason for rejecting it. However, some critics of the distinction have objected to it on the further ground that every scientific theory determines the observation predicates to be used in verifying the theory—that is, the observable data serving as tests of the theory are allegedly interpreted and formulated within a framework of assumptions that form part of the theory being tested.[8] This more radical thesis will now be briefly examined.

[8] See, for example, Feyerabend, "Problems of Empiricism," p. 214.

On the face of it, if this thesis were correct, arguments for accepting any theory on the basis of empirical data so manufactured would be fatally circular, since nothing would then count as pertinent evidence for the theory which failed to satisfy the relational patterns postulated by the theory. For example, if a moving body could properly be characterized as having a uniform velocity only if no external forces are acting on the body, Newton's first law of motion could never be in conflict with empirical findings, though at the cost of having no factual content whatsoever. However, the history of science provides numerous instances of theories that have been refuted by observational findings. In consequence, the claim that experimental data are always so selected or molded as to fit some assumed theory must be judged as untenable.

This difficulty in the thesis under discussion has been repeatedly noted, and some critics of the observation–theory distinction have tried to meet it. For example, Miss Hesse believes that an important part of the thesis can be saved, if two sorts of terms that could occur in observation statements are distinguished: terms presupposing the *full* truth of the theory being tested, from terms presupposing the truth of *only some* of the laws making up the theory. She acknowledges that if the terms formulating the evidential data for a theory are of the first kind, a genuine test of the theory is impossible; but she holds that if terms of the second sort are used in such formulations, circular reasoning of the type illustrated above does not occur in the test. For example, the truth as well as the meaning of the observation report that a body is moving with uniform speed in a straight line, is said to "depend . . . on the truth of laws relating measuring rods and clocks, and ultimately on the physical truth of the postulates of Euclidean geometry, and possibly of classical optics," all these laws being "part of the theory of Newtonian dynamics." But it is also claimed that since the evidential statement contains only predicates of the second kind, it can serve to test Newtonian theory without circularity.

This contention is patently correct, but the example (and the argument it illustrates) does not confirm anything of moment in the thesis under discussion. It is certainly possible to count the various laws concerning measuring instruments, as well as the laws of Euclidean geometry and even of optics, as parts of Newtonian dynamics. However, these laws do not constitute the distinctive content of Newtonian mechanics, nor is it these laws that are being tested by the observation report mentioned in the example. Thus, to take one instance, Euclidean geometry is known to be compatible with any number of non-Newtonian dynamical systems—the physical laws of Euclid do not entail any of the characteristic Newtonian laws (such as the first law of motion), nor are the meanings of the descriptive predicates occurring in the former determined by Newtonian assumptions. Since it is not Euclidean geometry which is presumably being put to a proof in the example, but rather a law specific to Newtonian mechanics and logically independent of Euclid, it is hardly surprising that no circularity is involved in the test. Moreover, the argument is incompatible with the view, integral to the radical thesis being discussed, that if a theory is changed or completely replaced by another, all observation terms used to state the evidential data for it necessarily change their meanings also. For the argument is built on the premise that some observation predicates used to state the evidence for a theory do not presuppose the truth of *all* the laws making up the theory. Such predicates can therefore continue to serve as observation terms with unchanged connotations for a different theory, obtained from the original one by changing one or more laws of the latter that are not presupposed by the predicates. In short, the argument makes coherent sense only on the assumption—associated with, but not distinctive of, the observation–theory distinction—that though the meanings of observation terms are at least partly determined by the laws into which the terms enter, the laws do not form a single, monolithic system of statements no two of which are logically independent.

This assumption also underlies Miss Hesse's attempt to counter a further objection to the thesis under consideration: if the meanings of the descriptive terms in observation reports were indeed fixed by the theory for which the reports can serve as evidence, the very *same* reports could not confirm each of two fundamentally *different* theories, nor confirm one and disconfirm the other. But if this were so, it would be impossible in principle ever to decide between ostensibly competing theories on the basis of observational findings—a conclusion that runs counter to actual scientific practice. But according to the proposed answer to the objection, this conclusion follows only in certain cases. If two theories really have no concepts in common, the conclusion is indeed inescapable. However, even theories that differ profoundly in their foundational assumptions and their implications may nevertheless contain "some hard core predicates and laws which they both share." For example, despite the far-reaching differences between

Newtonian and Einsteinian dynamics, both employ such important predicates as "acceleration of bodies falling near the earth's surface" and "velocity of light transmitted from the sun to the earth," and they also have in common a number of laws into which these predicates enter. In consequence, so the argument continues, if an observation report contains predicates occurring in laws that belong to both theories, the same report can be used to decide between the different theories. But here again the proposed resolution of the difficulty must grant the crucial point, vigorously denied by some critics of the observation–theory distinction, that the meanings of observation predicates are not completely determined by some given theory for which the predicates serve in formulations of evidence. The argument makes sense only on the assumption that though the meaning of a predicate P may be fixed (perhaps only in part) by some law L that belongs to each of two different theories T_1 and T_2, the meaning of P cannot depend on every other law in T_1 or every other law in T_2 if an observation report containing P is to test both theories. For otherwise, what the report would be saying when testing T_1 is different from what it would be saying when testing T_2, so that the *same* report would not (and indeed could not) be used to help decide between the two theories.

* * *

Science has often been compared with myth. Scientific theories, like many myths, are attempts to account for what happens in various sectors of nature; and, also like myths, they are works of the imagination that carry the impress of an enduring human condition as well as of variable special circumstances. It is therefore not surprising that scientific theories and myths have many common traits. Indeed, as Heinrich Hertz suggested in so many words in his treatise on the principles of mechanics, every system of symbols that makes up a scientific theory is bound to have components which can appropriately be called "mythical." For even when a theory is adequate to the facts for whose explanation it was devised, it will inevitably possess features that represent nothing in the subject matter, but reveal instead something about the capacities and preconceptions of its creator.

However, a controlling assumption underlying the development of science since ancient times is that properly conducted inquiry can arrive at explanatory theories that are not wholly mythical. If this assumption is sound, science and myth must not only be compared, but also contrasted. Such contrast has been frequently made: for example, by Einstein in the passage quoted at the outset, or by George Santayana when he declared that science differs from myth "insofar as science is capable of verification." To be sure, the probative value of verification has been seriously disputed throughout the centuries. And if the doubts about it raised by some recent commentators on the logic of science are well-founded, the belief that inspired the creators of modern science—that the truth about things can be found out by inquiry—is itself a myth. The critique in this paper of some currently influential arguments eventuating in a sceptical relativism is intended to show that those doubts are not warranted, that this ancient belief is still tenable, and that scientific theories cannot be equated on principle with old wives' tales.

Sylvain Bromberger

Science and the Forms of Ignorance

1

I am a very ignorant person and I have long assiduously deplored this fact. Ignorance is an invitation to scorn; it presumably deprives one of all sorts of joys, it can't be put to any useful ends, it can't be eaten, it can't be given away, and it is damn difficult to get rid of in any other way. Only those who are totally ignorant about ignorance can believe that ignorance is bliss! But ignorance can also define and determine the value of scientific contributions. Its study should therefore be one aspect of the study of how scientific contributions can be assessed, and it should thereby become part of the philosophy of science. This fact, if it is a fact, holds a heartening hope for someone like me. It means that a cause of shame and embarrassment may yet turn out to be a source of professional expertise! My talk should be followed in the light of that hope. Dale Carnegie, that great existentialist, somewhere quotes Robert Maynard Hutchins, that great essentialist, in turn quoting Julius Rosenwald, that great former president of Sears and Roebuck, as saying: "When you have a lemon, make a lemonade." This is what I propose to do here, and I invite you to share the refreshment, sour though it may turn out to be.

2

The point of the sort of study that I have in mind will be more easily seen by recalling some developments in the contemporary history of the philosophy of science. Until recently, most prominent philosophers of science centered their professional attention on a conception of scientific theories aptly and succinctly described in a recent paper by Hempel:

> The . . . conception of a theory as consisting of, or being decomposable into two principal components: (1) An uninterpreted deductive system, usually thought of as an axiomatized calculus *C*, whose postulates correspond to the basic principles of the theory and provide implicit definitions for its constitutive terms; (2) A set *R* of statements that assign empirical meaning to the terms and the sentences of *C* by linking them to potential observational or experimental findings and thus interpreting them, ultimately in terms of some observational vocabulary."[1]

Though the facts no longer warrant this, I shall refer to this conception as the *accepted view*.

It is important to remember that the proponents of the accepted view did not (or in any case need not and should not) claim that this conception correctly described what the word "theory" means or that it described the form in which theories occur in the literature. Either claim would have been blatantly false. The

[1] Carl G. Hempel, "The Structure of Scientific Theories," in Ronald Suter, ed. *The Isenberg Memorial Lectures, 1965–66* (East Lansing: Michigan State University Press, 1967).

claim was of a different nature altogether and might be put as follows: A theory is something that can be presented in many forms depending on one's objectives, one's audience, one's stylistic and pedagogic preferences, and other pragmatic considerations. However, every theory *can* be presented in the form of an abstract calculus paired with a set of interpretation rules satisfying the accepted view. Furthermore, the possibility of representing any theory in that form is of great philosophic significance. Why was it thought to be of great philosophic significance? The philosophers who developed the accepted view were committed to the task of making explicit the types of considerations that govern—or that ought to govern—the acceptance and the assessment of any alleged contribution to scientific knowledge, and, in particular, to the corpus of scientific theories. They also believed that a set of general criteria could be brought forth which (a) could be stated in ways that raised no philosophic difficulties, (b) would be unambiguously applicable to specific theories only in their guise of uninterpreted calculi paired with interpretation rules, (c) would rely in an essential way on the properties of calculi and interpretation rules, (d) would conform to the tenets of empiricism and the interests of enlightened scientists, (e) would mark as acceptable all and only those theories that scientists and empiricists would admit as acceptable on the basis of pre-analytic intuitions and philosophic conscience. Thus the adherents of the view saw it as indicating how their task should be approached and completed. If a theory can always be represented as an uninterpreted calculus cum interpretation rules, and if criteria exist that satisfy (a), (b), (c), (d), and (e) above, then, in a sense that is difficult to analyze but easy to grasp, the task becomes "simply" that of formulating these criteria.

Those were not the only considerations that encouraged investment of time and thought and reputation in the accepted view. There are a number of concepts used in talking *about* theories—and particularly when evaluating them—that have long puzzled analytic philosophers. They include the notions of explanation, of theoretical terms, of predictive power, of empirical content, and others. Assuming the accepted view to be right, it seemed reasonable to look forward to an explication of these notions (in terms of the unproblematic notions that characterize interpreted calculi) that would constitute a solution of the analytic puzzles.

The accepted view is less and less accepted. Why? Some reasons for the decline are purely sociological. But there are less earthly reasons as well. The accepted view entailed at least four claims: (1) that every scientific theory can be axiomatized, i.e., cast in the form of an interpreted calculus fitting Hempel's above description; (2) that a set of criteria can be stated that meets conditions (a) to (e) above; (3) that every accepted theory, once axiomatized, will satisfy these criteria; (4) that the notions of explanation, theoretical term, etc., can be reduced to the notions needed to formulate the accepted view. Three of these claims, i.e., (1), (2), and (4), entail the possibility of carrying out certain programs. These programs have not been carried out so far, and obstacles that appear insurmountable have arisen in their path. Thus the failure of the programs so far constitutes grounds for skepticism, i.e., grounds for believing that there is something essentially mistaken about the accepted view.

I believe that the skeptics are right and that the accepted view is no longer tenable. But to say this is *not* to say that the problem that motivated it should also be dismissed. Remember the problem: to find and to state in philosophically unobjectionable ways the principles that govern or that ought to govern the acceptability of any alleged contribution to science, and especially to the corpus of theories. Let us call this the Appreciation Problem. Many people now deem also that the Appreciation Problem itself rests on a mistaken presupposition, viz. the presupposition that there are invariant, universal principles of acceptability which survive the continuous overthrow of scientific doctrines and which regulate their replacement. I do not wish to show here that the objections against that presupposition miss their mark. Nor am I prepared to defend the presupposition. I view it as an article of philosophic faith, supported by some clear intuitions, but possibly false. The only way to find out is by trying to solve the Appreciation Problem.

3

If I am right, we need a new approach to the Appreciation Problem. How shall we proceed? Let me first draw on my expertise as a master ignoramus to make a few initial suggestions that I shall refine somewhat later.

How do I know that I am such a good ignoramus? Well, I can cite a number of questions whose answers I don't know: "what is the distance between London and Paris?" "how many arithmetic steps are absolutely required in a program for solving a set of linear equations?" "what is the Papago word for 'horse'?" "what is the atomic weight of calcium?" Furthermore, some questions have me so utterly baffled that I can't even produce a plausible answer, e.g., "why do tea kettles emit a humming noise just before the water in them begins to boil?" "what are the heuristics by which a child discovers the grammar of his language?" Finally, there are problems that have me stumped. Give me the length of a ship, the weight of its cargo, the height of its mast, and I still can't figure out the age of the captain. Some people can. Or perhaps they can't do that, but give them the percentage of a solution of ethylene glycol in water and its density and they will figure out for you the vapor pressure of that solution at 20°C. In short, I know questions whose correct answers I can't tell from incorrect ones; I know questions to which I know only incorrect answers; I know problems that I can't solve.

But this now suggests that my ignorance is not one big undifferentiated glop, one huge unstructured nothingness. It is apparently made up of identifiable items. These items furthermore lend themselves to classification and comparison. It even looks as if they may stand in interesting relationships to one another and to other things. Let us tentatively refer to such items as *items of nescience.*

So far I have talked as if these items of nescience were my personal property. This is obviously presumptuous. Many are the common possession of a number of us and some are shared by all of us. It is therefore possible to think of them in abstraction of any specific owners. Thus a question that is an item of nescience of anyone who does not know the answer can be thought of as simply an item of nescience, without regard to its actual status in this or that person's ignorabilia.

Two more pieces of loose—but I hope suggestive—

jargon will help to expedite matters. Let *p* be a correct answer to some question *q*. No one can count *q* among his items of nescience who knows that *p*. I shall say in that case that *p cancels q*. For instance "Dante was born in 1265 in Florence" cancels "Where and when was Dante born?" Questions, as we shall see, are not the only type of items of nescience. Shortly, I shall describe other types and extend the concept of cancellation to cover them. Let *x* now be an item that cancels *some* item of nescience. I'll then refer to *x* as an *item of yescience*.

With this last ugly new expression I can now state a philosophical hypothesis: Science is a set of actual or presumed items of yescience. It is no doubt a strange hypothesis. On the one hand, it looks like a truism; on the other, it hardly makes sense. The fact that it looks like a truism need not deter us. On the contrary! After all, philosophy consists in taking truisms seriously and squeezing depth out of them. The fact that it barely makes sense need not deter us either. It simply shows that we have some analytic work cut out for us. The important point to notice about the hypothesis is that it offers a new construal of the Appreciation Problem, for that problem can now be stated as follows: what principles determine whether a presumptive contribution to science is an item, or a set of items, of yescience? i.e., what determines whether it cancels what item of nescience, if any?

Is the hypothesis true? Well, it is too vague at this stage to enable us to decide. However, one thing must be clear: the hypothesis does not assert that every actual contribution or alleged contribution to science was initially sought in response to an awareness that something specifiable as an item of nescience needed to be cancelled. There are innumerable contributions of which this is true. But there are many of which it is not. The truth of the hypothesis then should not hinge on the presence or absence of certain episodes in the intellectual biography of scientists. Furthermore, the test of the hypothesis lies in its entailments, and, particularly, in its programmatic entailment. In this respect it resembles the accepted view. I shall turn to that program in a moment. It constitutes an approach to the Appreciation Problem. If it can be carried out, this will constitute grounds for retaining the hypothesis, and for deeming that the Appreciation Problem can be solved. If it can't, then the hypothesis will have to be rejected and the Problem will have to be recast once more.

4

Before going any further let us glance a litttle more closely at some of the flora and fauna of the land of nescience and yescience. I have mentioned questions and answers. "Nixon is the President of the United States" cancels unfortunately "who is the President of the United States?" But now consider "how is speech produced?" or "what heuristics do good chess players use to spot possibilities?" or again "what would the universe be like if nothing existed?" We are, with regard to these questions, in what I have called elsewhere[2] a *p-predicament* ('p' here stands for 'puzzled' but as mnemonic aid only), a predicament that I have described as fol-

[2] On the notion of p-predicament, see my "An Approach to Explanation" in *Analytical Philosophy*, 2nd Ser., R. J. Butler, ed. (Oxford, 1965). Also my "Why-questions" in *Mind and Cosmos*, University of Pittsburgh Series in the Philosophy of Science, vol. III (Pittsburgh: University of Pittsburgh Press, 1966).

lows: someone is in a p-predicament with regard to a question *q* if and only if, given that person's assumption and knowledge: *q* must be sound, i.e., must have only true presuppositions (and hence have at least one correct answer), yet that person can think of no answer, can imagine no answer, can conjure up no answer, can invent no answer, can remember no answer to which, on his view, there are no decisive objections. To cancel a p-predicament, less than a correct answer is required. Any answer not known to be false will do, even if not known to be true. It often requires the greatest geniuses to discover some answers of that sort. Consider also "why does the moon look larger when near the horizon?" If my information about that question is right, we do not know the answer, but a number of still viable candidate answers are available. We are not in a p-predicament with regard to the question. Nevertheless, we also know that there probably are many other possible answers that have not occurred to anyone yet. The discovery of such further possibilities would enhance our science and would thus also cancel some nescience. Answers that cancel items of nescience of the sort illustrated with the aid of the last four questions are what I call *theories*. I have given an explicit definition of that term elsewhere.[3] Professor Achinstein, one of the very few people to have looked at that definition, has been kind enough to take a few whacks at it.[4] I believe that I can easily quibble my way out

[3] See my "A Theory about the Theory of Theory and about the Theory of Theories" in *Philosophy of Science, The Delaware Seminar*, William L. Reese, ed. (New York: 1963).

[4] Peter Achinstein, *Concepts of Science* (Baltimore: Johns Hopkins Press, 1968), pp. 134–47.

of his criticisms, but they do bring out certain very unfortunate unclarities that discourage me from repeating the definition here. I'll reintroduce the term later in the discussion.

If you wish to know the distance between Baltimore and Boston, you might get yourself a yardstick and measure the distance. It is a very tedious way of getting an answer to "what is the distance between Baltimore and Boston?" and it may even be an impossible one. You can, however, get the answer by measuring the distance between the points corresponding to these cities on a good map and by then computing the real distance from that one with the aid of the scale. Or again, if you wish to know the distance traveled by a free falling body in half an hour, you can get yourselves a big vacuum and drop objects from various heights until you find one that requires thirty minutes. This too would be tedious and probably impractical. But you can get the answer to "what is the distance traveled by a free falling body in half an hour?" simply by substituting '30' in the version of '$\frac{1}{2}$ gt^2' that has the parameters in c.g.s. units. In each of these cases, there are thus at least two ways of arriving at an answer, a direct method and one that derives the answer through computations on answers to other questions. Let us call the latter *procedures for carrying out vicarious investigations,* and the conglomerate they make up, we shall call *theory* (not *theories*; the word here is used as a mass-noun as in *electromagnetic theory* and no longer as a count-noun as in the *theory of relativity*). I mentioned earlier that if you gave me the length of the ship, the weight of the cargo, the height of the mast, I still would not be able to figure out the age of the captain. I could, of course, ask him.

But there is, as far as I know, no available method for carrying out a vicarious investigation that will yield an answer to the question about the age of the captain and that will rely solely on the information given in the problem. Our item of nescience here is the lack of a formula for computing the answer from that information, and if it can be cancelled, it will be cancelled by such a formula, or by a method for deriving such a formula from general principles.

As we can see, items of nescience and items of ye-science belong to very different species. We could, of course, say that there are only questions and answers, and that in the allegedly other sorts of cases we simply have slightly complex questions, e.g., "what is an answer neither known to be true nor known to be false to 'how is speech produced?'?" or "what would be a method for carrying out a vicarious investigation using the answers to 'what is the length of the ship?,' 'what is the weight of the cargo?,' 'what is the height of the mast?' to obtain the answer to 'what is the age of the captain?'?". We could say this, but we would then blur distinctions that are important in our approach to the Appreciation Problem. Since I do not wish to blur any more distinctions than I shall be compelled to blur anyhow, I shall pretend throughout that we do not ask questions about questions, except when we do philosophy.

5

Back now to the Appreciation Problem. You will recall its new construal: what principles determine whether a presumptive contribution to science is an item or a set of items of yescience, i.e., what determines whether it cancels what items of nescience?

To solve that problem we need a general theory of nescience and yescience that will exhibit such principles in a form applicable to actual cases (though these may themselves have to be put first in some canonical form—as was the case for the accepted view) and satisfy ourselves, perhaps by looking at actual cases, that these principles yield the very evaluations we make when flying by the seat of our rational pants.

This is an enormous order which I cannot hope to fill in this talk or even in my life. I can and will indicate the broad outlines of that theory and thus describe the program that will test the hypothesis I am entertaining. I shall do this by outlining the chapters of an imaginary treatise through which the theory of nescience and yescience might be taught. This treatise will contain at least three chapters: I. Erotetic Logic, II. Theory of Theories, III. Theory of Theory. Each chapter will in turn be divided in two sections, a deductive section and an inductive section. What will these chapters and sections contain?

I. Erotetic Logic

Erotetic logic is the logic of questions and answers. This chapter starts from the assumption that every question stands in three different relations to specifiable propositions. Some propositions *give rise* to it; some propositions are *presupposed* by it; some propositions are *direct answers* to it. For instance, "the Empire State Building is heavy" gives rise to "how heavy is the Empire State Building?". "There is a King of France" is presupposed by "what is the age of the present King of France?". "The present King of France weighs five hundred pounds" is a direct answer to "what is the weight of the present King of France?" (So is "the present King of France weighs two hundred pounds.") Roughly put, a proposition *gives rise* to a question if it entails that the question has a correct answer. A proposition is a *presupposition* of a question if its falsehood entails that the question has no correct answer. A proposition is a *direct* answer to a question if, if true, it is a correct answer.

Notice that if a proposition gives rise to a question, then it entails the presuppositions of that question. The converse does not hold. Notice also that every direct answer (whether correct or not) entails the presuppositions of the question, but does not necessarily entail that which gives rise to that question.

The deductive section of erotetic logic concerns itself with all those relations. If the approach is formal, then this section will first develop ways of representing propositions and questions and it will then explicate the relations of giving rise, presupposing, and being a direct answer in terms of these methods of representation. In other words, it will develop methods of representation, and concepts about these methods of representation, and explications that together will satisfy the following conditions: if p gives rise to q, then the fact that p gives rise to q can be inferred from the form in which p and q are represented. If p is presupposed by q, then this too can be inferred from the forms of p and q as represented by our method. If p is a direct answer to q, then this again can be inferred from the forms. In all these respects a formal erotetic logic simply apes formal deductive logic. In fact, a wise approach will try to develop this chapter so that it can be amalgamated with some system of formal "assertoric" logic (we owe the term to Nuel Belnap). If the approach is not formal, then a theory of non-formal features of some sort will have to be developed which satisfies the condition that if *p* gives rise to *q* then it will turn out that *p* and *q* share some features common to all and only pairs which stand in that relation. Analogous non-formal features will have to be found for the other two relations.[5]

So much for the deductive section of erotetic logic. The inductive section concerns itself with the relation between a question and direct answers deemed correct or probable, or at least warranted by evidence. If this part is formal too, then the methods of representation will have to be such that if *q* is a question and *R* is a set of propositions which entail that *p* is a correct (or probable or warranted) direct answer, then this fact will have to be inferable from the form of their representation. Parts of the inductive section will have to wait for developments in the theory of theory, as will become evident shortly. Furthermore, this section will also have to include most of the traditional domain of inductive logic since it concerns itself with the grounds for assigning truth, or probability, or warrantedness to propositions that give rise to or are presuppositions of questions, as well as to those that express answers.

So far I have not used the words "item of nescience" or "item of yescience" or mentioned the Appreciation Problem. Questions are genuine items of nescience if

[5] References to the systems of erotetic logic will be most conveniently found by consulting the footnotes of David Harrah's paper cited below.

and only if their presuppositions are true. A question is a spurious item of nescience if at least one of its presuppositions is false. Any proposition that gives rise to a question generates, if true, an item of nescience. Nothing can cancel a question unless (a) it is a direct answer to that question and (b) the question is not a spurious item of nescience. A question is cancelled when one of its direct answers is confirmed. It is tentatively cancelled when one of its direct answers is shown to be probable or warranted. The Appreciation Problem requires that we work out the conditions under which any of these contingencies hold, and the chapter we have not written just now informs us about these conditions.

II. The Theory of Theories

The theory of theories concerns itself with p-predicaments and their cancellation. A person was said to be in a p-predicament with regard to a question *q*, you will recall, if, given that person's knowledge and assumptions, *q* must be sound, i.e., have no false presuppositions, yet the person can think of no answer, can conjure up no answer to which, given that knowledge and those assumptions (and nothing else, I should add), there are no decisive objections. Futher, a p-predicament with regard to a question is cancelled by any direct answer compatible with that knowledge and those assumptions. But the theory concerns itself also with the cancellation achieved by the discovery of new compatible answers, when some compatible ones are already known. In view of this, the discussion can be simplified if we modify the characterization of p-predicaments slightly to cover all the relevant cases. Let us then say that a person is in a p-predicament with regard to a question *q* if there is *some* answer compatible with his knowledge and assumptions that he cannot think of, cannot excogitate, etc. Under this new description one need not be as much in the dark to be in a p-predicament: one may have conjured up some direct answers that are compatible with one's set of beliefs; one may even know some correct answers. I, for instance, am now in a p-predicament with regard to "why did that cup I dropped this morning break?" even though I can think of "that cup broke because it was brittle and hit a hard floor," for there are many other answers, some of which will undoubtedly occur to physicists, that do not occur to me.

The deductive part of the theory of theories must explicate the notion of p-predicament. To accomplish this it will have to incorporate and build on many parts of erotetic logic. The systems of erotetic logic presently available, i.e., those of Kubinski, Harrah, Belnap, and Åquist, all use relations of direct answerhood that are *constructive*. In other words, in each of these systems the whole set of direct answers to a question can be generated by an algorithm. Offhand it may seem that in order to represent adequately the notion of p-predicament the theory of theories should rely on a system of which this is not true. However, Harrah has recently shown[6] that systems that incorporate such algorithms are *incomplete* in the non-pejorative sense that they entail the existence of sets of propositions that are

[6] David Harrah, "On Completeness in the Logic of Questions," *American Philosophical Quarterly* 6, no. 2 (April, 1969).

not the set of direct answers of *any* question. The proof is very simple and need not detain us. The theorem itself accords with intuition, i.e., seems to be true not only of artificial systems, but also of the ones embodied in us. For this reason, as well as because of some other premonitions—not to be mentioned here—about the eventual character of an adequate erotetic logic (the systems presently available are simply not rich enough), I assume that in the system of our theory the relation of direct answerhood will have to be constructive. The algorithms however will in many cases turn out to be either not obvious or not simple or not manageable within human limitations. But I complement this assumption with the further assumption that we will be able to define *search algorithms,* i.e., algorithms that enumerate proper subsets of the set of direct answers of some questions and that express heuristics for the excogitation of possible answers.

A p-predicament can then be thought of as involving five elements: (1) a person, (2) a set of propositions expressing the knowledge or assumptions of that person, (3) a question whose set of direct answers can, in principle, be generated by an algorithm, (4) a finite (possibly empty) set of direct answers to the question available to the person and compatible with the propositions in the above set, (5) a search algorithm that generates only direct answers incompatible with the set of propositions in (2)—unless already members of the finite set of direct answers in (4)—and that expresses the heuristic the person would use (consciously or unconsciously) in seeking possible direct answers. Since we are dealing with a philosophic theory, our purposes will be best served by forgetting persons. Let us then define a p-predicament as an ordered quadruple consisting of a set of propositions, a question, a finite (possibly empty) set of direct answers to that question, and a search algorithm. I shall refer to the set of propositions as *the constraint* and require that the constraint never include propositions that give rise to the question. This will simplify the discussion somewhat. The search algorithm must only generate such direct answers to the question as are either members of the finite set of direct answers or are incompatible with members of the constraint. This condition will some day be supplemented with conditions on the algorithms which generate all the direct answers. These conditions will rule out cases where such algorithms already represent obvious heuristics.

Each p-predicament, i.e., each such ordered quadruple, is an item of nescience. What of the question in it? It, too, is an item of nescience. Each p-predicament represents then two items of nescience: itself, and the question in its bosom. The difference between them lies in their structure and in the manner in which they are cancelled. The p-predicament (unlike the question) is cancelled by any direct answer (to the question) that is not produced by its search algorithm and that is not in its finite set of direct answers, and that is compatible with the members of the constraint.

The deductive part of the theory of theories explicates all these notions along lines similar to those in the corresponding section of erotetic logic. If the theory is formal, then it will incorporate representations with the power to reveal whether a given quadruple is (or fails to be) a p-predicament from the way the elements are represented. These methods of representation will have the same power for the relation of cancellation. If the theory is not formal, then it will reveal non-formal features shared by all and only p-predicaments, and non-formal features shared by all and only pairs consisting of a p-predicament and an item of yescience that cancels it.

The inductive section of the theory of theories repeats most of the inductive section of the previous chapter, but it adds some elements of its own. We shall only glimpse these here. First two analytic definitions.

A proposition s is an *explanation relative to a p-predicament* x if (a) s is true and (b) at least one of the following conditions holds: (i) the members of x's constraint are true, the presuppositions of the question in x are true, and s cancels x; or (ii) the members of x's constraint are true, some of the presuppositions of the question in x are false, and s is a denial of these false presuppositions; or (iii) some of the members of x's constraint are false, the presuppositions of the question in x are true, and s is a conjunction of an answer to the question in x and of denial of the false constraints incompatible with that answer.

A proposition t is a *theory relative to a p-predicament* x if (a) t is neither known to be true nor known to be false and (b) at least one of the following conditions holds: (i) t cancels x; or (ii) t is the denial of some presuppositions of the question in x; or (iii) t is a conjunction of a direct answer to the question in x and the denial of members of x's constraint incompatible with that answer. Roughly then, a theory is a possible explanation in that it is an explanation if true.

I shall not defend these definitions here. Their rationale, if there is any, can readily be extracted from my other papers on this subject.[7] The inductive section deals with both explanations and theories, but especially the latter. It concerns itself with the conditions on data, p-predicaments and propositions (remember that direct answers are also propositions) which determine when the latter are theories and, if they are, relative to what p-predicament and to what body of data. The approach here again may turn out to be either formal or non-formal. The inductive section also concerns itself with the conditions under which a theory becomes a warranted proposition by virtue of being the only theory available relative to a specific p-predicament.

The details of this section are bound to be gruesome. When available, they will constitute another part of the solution of the Appreciation Problem.

III. The Theory of Theory

The theory of theory concerns itself with procedures for the conduct of vicarious investigations. The topic is vast and has been outlined in some detail elsewhere.[8] Brief suggestions will have to serve here. Remember our examples: the measuring of a distance between two cities, and the determination of a certain time interval followed by the computation therefrom of the distance traveled during that interval by a free-falling body. In each case there was a question whose answer we wanted, there were questions whose answers we knew or assumed, and there was a computation that led

[7] See my "Approach to Explanation"; "Why-questions"; and "A Theory about the Theory . . . ," cited above.

[8] *Ibid.*

from the latter answers to the former ones. A problem as envisaged here is thus representable as a set of questions, some of which are marked as answered and one of which is marked as unanswered. As an item of nescience it is cancelled by a partial recursive function which maps some direct answers of the questions marked as answered on some direct answers of the question marked as unanswered. The partial recursive function must, however, map correct answers on only correct answers. But problems fall into certain natural types. Thus the formula used to compute the distance traveled by our free-falling body could have been used to solve *any* problem whose unanswered question was about a distance traveled by a free-falling body and whose answered one was about the duration of the fall. Even without knowing the formula we could have predicted from the similarities of these problems that there should be one formula for all of them. Problem types then also constitute items of nescience. They are cancelled when a partial recursive function is found that will map as above for any member of the type.

The deductive section of the theory of theory explicates the notions of problems, problem types, and of their cancellation. For instance, it offers ways of representing (or features of) functions and problems which determine whether a given function maps direct answers marked as known on answers marked as unknown in a given problem. It also presents ways of determining whether a set of problems constitutes a type.

The inductive section of the theory of theory investigates the conditions under which formulae (i.e., the sort of partial recursive functions discussed above) can be trusted to map true, or probable, or warranted answers on true, or probable, or warranted ones in a problem. It also investigates when an answer can be deemed warranted *by virtue of the fact* that it is the value of a formula for certain known answers of a given problem. It is thereby also relevant to the inductive section of erotetic logic.

6

Composing the treatise that I have just outlined will be only one half of the program required to solve the Appreciation Problem and to show that science is a set of actual or presumed items of yescience. The other half requires that we see whether the actual corpus of admittedly scientific doctrines conforms to the theory. We will therefore have to express contents of that corpus so as to reveal them as either confirming or disconfirming instances.

Is there any reason to believe that either half can be accomplished? The question is vacuous until we impose constraints on what devices, what concepts, what symbols may enter into the theory of nescience and yescience. But what would be reasonable and plausible constraints? I would like to suggest the following, at least for the deductive sections: that the devices be essentially those also required in an elaboration of the syntax and semantics of *natural* languages, but an elaboration that presents explicitly aspects of grammars and of semantics internalized by human beings. A constraint on a philosophic theory is reasonable and plausible when it reflects a reasonable and plausible point of view. The constraint that I have suggested would reflect the view that expressibility in a human language is of the essence in scientific knowledge, i.e., that nothing can constitute an addition to our scientific knowledge unless it can be stated in some language or other, some *human* language or other, and thus be taught and shared, and that nothing counts as an instance of ignorance amenable to scientific research unless it too can be conveyed in a human language and jointly recognized.

This may seem to mean that we must postpone our program until further progress is made by linguists. But not really. Enough is known or can be guessed to enable us to draw tentative sketches and to peruse the works of scientists for plausible items of yescience. Anyone can take a text and ask himself: what questions are answered here, what theories are proposed, what methods for carrying out vicarious investigations are offered? After a while, patterns begin to emerge and generalizations suggest themselves that can be put in terms that we also know how to apply to our language. This already gives some prima facie support to our hypothesis. It can also be quite exhilarating. It brings the works of scientists alive in a way that is not matched by any other philosophic perception. Karl Popper invites us to look upon scientific progress as a series of combats in which each contribution is either a chip on someone's shoulder or an attempt to knock off a chip. Thomas Kuhn invites us to think of scientific progress as exercises in imitation interrupted by changes in fashion. The view proposed here sees scientific progress as something more worthy of emulation by other human endeavors, as work designed to overcome identifiable shortcomings shared by many, as production for consumption, or, more simply, as one aspect of the eradication of ignorance.

Adolf Grünbaum

Can We Ascertain the Falsity of a Scientific Hypothesis?

1. Introduction

Can we ascertain the falsity of a given scientific hypothesis H? Alternatively, could we ascertain the truth of H? One tradition answers these questions *asymmetrically* as follows: Alas, unfavorable results furnished by just one kind of experiment suffice to guarantee the falsity of an otherwise highly successful hypothesis. And would that *favorable experimental* findings had a comparable capability of establishing the *truth* of a hypothesis! Thus, the scientist is held to be laboring under a discouraging handicap in his quest to glean nature's secrets. His most triumphant theories are never safe from refutation by potential contrary evidence. Hence, none of his hypotheses can ever be known to be true with certainty. But if even a small amount of contrary evidence does materialize, then the most celebrated of hypotheses is indeed known to be false.

On this view, the asymmetry between provability and *dis*provability arises from two simple facts of elementary deductive logic. If an observational consequence O of a hypothesis H turns out to be true, then the truth of H does *not* follow deductively. The reason is that the truth of H is *not* necessary for the truth of O but only sufficient. To infer the truth of H deductively here would be to commit the so-called fallacy of affirming the consequent. Yet, if alternatively the observational findings do *not* accord with O, then the falsity of O does entail the falsity of H. For if H were true, *all* of its observational consequences, including O, would have to be true. The deduction of the falsity of H in this case is held to be valid by conforming to the so-called *modus tollens* form of valid inference.

This asymmetrical account of the verifiability vis-à-vis the falsifiability of a hypothesis leads to the claim that there are CRUCIAL *experiments* in science. For suppose that two rival hypotheses H_1 and H_2 each explain a given large set of observational facts. But suppose further that these hypotheses entail incompatible observational results O_1 and O_2 respectively as outcomes of a given kind of experiment C. And let the performance of C (on one or more occasions) issue in the observation O_1. In that case, we are told, C is a crucial experiment because it decisively eliminates H_2 from the scientific arena as conclusively refuted by O_1. Therefore, the experiment C confirms H_1 vis-à-vis H_2.

But this confirmation of H_1 vis-à-vis H_2 is not at all tantamount to a conclusive verification of H_1 per se. After all, the outcome O_1 of C cannot render H_1 immune to refutation by potential contrary evidence O_3. Indeed, the favorable outcome of C cannot assure the success of H_1 in a potential confrontation with another rival hypothesis H_3 in some other crucial experiment C′. In short, according to the asymmetrical account, there are crucially falsifying experiments but not crucially verifying ones.

The many treatises and textbooks of science which characterize certain experiments as crucial attest the influence of this account. For example, the experiment which Galileo allegedly performed from the Leaning Tower of Pisa is often presented as a decisive refutation of Aristotle's hypothesis that the acceleration of a heavy falling body is greater than that of a lighter one. And in a recent encyclopedia article, the renowned geneticist G. W. Beadle credits Louis Pasteur's experiments of 1862 with being crucial. Says he: "Louis Pasteur completely *disproved* the theory of spontaneous generation. He showed that bacteria would not grow in materials which were sterilized" (my italics).[1]

But if Pasteur's experiments of 1862 did indeed constitute a complete disproof of the hypothesis of spontaneous generation of life, as claimed here by Beadle, then one wonders at once how that hypothesis could have been rehabilitated by A. I. Oparin in 1938 and further by H. Urey in 1952 to the following effect: life on earth originated by spontaneous generation under favorable conditions prevailing sometime between 4.5 billion years ago and the time of the earliest fossil evidence 2.7 billion years ago. Indeed, the conception of crucial experiment exemplified by Beadle's overstatement is an oversimplification of the falsifiability of a scientific hypothesis.[2] To show this, I shall first consider briefly the logic underlying the purported disproof

[1] G. W. Beadle, "Spontaneous Generation," *World Book Encyclopedia* (Toronto: Field Enterprises Educational Corp., 1967), vol. 17, p. 626.

[2] In his *Induction and Intuition in Scientific Thought* [American Philosophical Society Memoir, vol. 75, Philadelphia, 1969], Peter B. Medawar gave a telling illustration from *cancer research* (p. 54, *n.* 44) of the logical pitfalls which can jeopardize the reasoning underlying purported crucial experiments

of the occurrence of spontaneous generation of life. And then I shall examine an important episode in the history of recent astronomy culminating in the claim that R. H. Dicke's observations of the flattening of the sun disprove Einstein's general theory of relativity.

2. Purported Disproofs of Hypotheses in Biology and Astronomy

1. The Biological Hypothesis of Spontaneous Generation

The spontaneous generation of life is sometimes called "abiogenesis." For the sake of brevity, I shall refer to the hypothesis of spontaneous generation as the "SG-hypothesis" or simply as "SG."

Presumably, SG states that living systems did and indeed would develop after an unspecified time interval under unspecified kinds of conditions which actually prevailed on the earth sometime during the past. But it plainly does *not* follow from *this* formulation of SG *alone* that *bacteria* would grow even once—let alone every time—in sterilized materials under the conditions of Pasteur's experiments during the time periods which he allotted to them. What then is the probative force of Pasteur's negative results as a basis for refuting SG? Let us suppose, just for argument's sake, that Pasteur's

pertaining to transformations of cells into the cancerous state. Karl Popper's influential version of the asymmetrical account of verifiability and falsifiability has very recently been criticized by B. Juhos, "Die methodologische Symmetrie von Verifikation und Falsifikation," *Zeitschrift für allgemeine Wissenschaftstheorie* I (1970): 41–70, especially 52ff.

own failure to obtain bacteria had established conclusively that there is no instance at all in which bacteria would grow spontaneously in his kind of experiment. Even then his experimental results could *not* have refuted the very general SG-hypothesis via *modus tollens.* The obvious reason is that a theory can be refuted by Pasteur's negative results *only* if that theory entails a positive or contrary outcome for his kind of experiment, which SG alone does not.

The entailment of a positive outcome could be assured by conjoining SG with a suitable auxiliary hypothesis *A* as follows: The auxiliary hypothesis *A* states that the time interval during which abiogenesis would occur need only be a small fraction of a man's life span, and furthermore *A* confers specificity on SG by supplying the kinds of physical conditions relevant to Pasteur's experiment, such as the presence of an oxidizing atmosphere.

Hence what Pasteur's negative results refute is *not* SG itself, but only the *conjunction* of SG with *A*. Indeed, SG can be upheld in the face of his findings by blaming the false prediction of a positive outcome of his experiments on the falsity of *A*. And *A* can therefore be replaced by a suitably different auxiliary hypothesis *A′* to confer specificity onto SG. No wonder, therefore, that Oparin, Urey, and others later felt free to postulate the truth of SG as part of a speculative, augmented theory. Their kind of theory asserted very roughly the following: When the earth was first formed, it had a reducing atmosphere of methane, ammonia, water, and hydrogen. Only at a later stage did photochemical splitting of water issue in an oxidizing atmosphere of carbon dioxide, nitrogen and oxygen. The action of electric discharges or of ultra-violet light on a mixture of methane, ammonia, water, and hydrogen yields simple organic compounds such as amino acids and urea, as shown by work done since 1953.[3] The first living organism originated by a series of non-biological steps from simple organic compounds which reacted to form structures of ever greater complexity until producing a structure that qualifies as living.

Commenting on this particular augmentation of SG, Beadle says that "no one has ever demonstrated or proved this development under satisfactorily controlled conditions."[4] But as long as future evidence may still sustain these speculations, as Beadle seems to allow here, then it is clearly too strong to claim that "Louis Pasteur completely disproved the theory of spontaneous generation."

2. *The General Theory of Relativity*

Let us now turn to the evaluation of a recent purported experimental disproof of the general theory of relativity. To set the stage for this appraisal, let us first

[3] S. L. Miller, "Life, Origin of," in *The International Encyclopedia of Science*, ed. James R. Newman (New York: Thomas Nelson & Sons, 1965), vol. II, p. 622. A. I. Oparin, *Genesis and Evolutionary Development of Life*, translated from the 1966 Russian edition (Moscow: Meditsina Publishing House) by Eleanor Maas (New York: Academic Press, 1968); also available as NASA Technical Translation TT F-88 (Washington: NASA, 1968). See also J. D. Bernal, *The Origin of Life* (Cleveland: World, 1967); D. H. Kenyon and G. Steinman, *Biological Predestination* (New York: McGraw-Hill, 1969); and P. Handler, ed., *Biology and the Future of Man* (New York: Oxford University Press, 1970), Chap. 5, "The Origins of Life."

[4] Beadle, "Spontaneous Generation."

have a very brief look at the logic of the threat posed to Newton's theory of motion and gravitation by two sets of observational findings. First, there was a discrepancy between the *observed* motion of the planet Uranus and its motion as *calculated* theoretically about 1820. This calculation was obtained from Newton's theory via Laplace's determination of the perturbations superposed by Jupiter and Saturn on the sun's gravitational attraction. The second threat to Newton's theory arose from the magnitude of the observed slow precession of Mercury's perihelion or of its longer orbital axis in the direction of its motion. The observed precession yielded a troublesome excess of about 43 seconds of arc per century. In Newton's theory, the orbit of a planet due to the sun's gravitational attraction alone would be a closed ellipse with the sun at one focus. Hence the Newtonian will seek to account for the slow rotation of Mercury's ellipse by means of perturbations emanating from the other planets, notably from Venus, which is the closest, and from Jupiter, which is the most massive. But such a Newtonian calculation fails to account for a residuum of 43 seconds of arc per century from the observed value of the precession. Let us see how Newton's theory attempted to deal with the challenge of these two sets of observations.

In the case of Uranus, the theory from which its orbit was calculated in about 1820 included the hypothesis H of Newton's laws and the auxiliary assumption *A* that the relevant perturbations emanated *entirely* from Jupiter and Saturn. Historically, the empirical incorrectness of the calculated orbit of Uranus was blamed *not* on the falsity of the Newtonian H but rather on the falsity of *A*. Specifically, Adams and Leverrier each theoretically postulated the previously unsuspected existence of the planet Neptune, thereby introducing a contrary auxiliary hypothesis A′. It was the first case of inferring a perturbing planet from the observed perturbations rather than conversely.[5] Ironically, Lalande had unwittingly observed Neptune as early as May 1795. Having assumed it to be a fixed star, he attributed the discordance between its apparent positions two days apart to observational errors and discounted at least one of these two observations.[6] It is interesting to note what probative significance was attributed to this detection of Neptune through its action on Uranus before Neptune's presence was corroborated observationally. The American astronomer, Simon Newcomb (1835–1909), hailed it as "a striking example of the precision reached by the theory of the celestial motions."[7] The theory to which Newcomb attributes this virtue here includes among its component premises Newton's H and an A′ which asserts the existence of *eight* of the now known nine planets.[8]

But this very combination of H and A′ yielded an observationally *incorrect* value for the precession of the perihelion of Mercury. The astronomers then made clear how difficult it is to disprove observationally once and for all any one component hypothesis of a total theoretical system. For they noted that the observed

[5] For details, see A. Pannekoek, *A History of Astronomy* (New York: Interscience Publishers, 1961), pp. 359–63.

[6] "Neptune," *The Encyclopædia Britannica*, 14th ed., 1929, vol. 16, p. 228.

[7] *Ibid.*, p. 226.

[8] Pluto was not inferred from the perturbations in Neptune's own orbit until 1931.

greater value does not enable us to pinpoint the blame as between the components H and A′ of the total celestial mechanics. Thus, assuming that only *one* of the two components H and A′ is false, Leverrier considered a new auxiliary hypothesis A″ which postulated a group of planets revolving inside the orbit of Mercury to account for the discrepancy. And *at first,* it even seemed that the existence of such intra-Mercurial bodies was confirmed by occasional reports of dark objects seen in transit across the sun's face. But numerous photographs failed to bear out Leverrier's A″.[9] Hence other astronomers allowed that Newton's H might be false. They modified Newton's law of gravitation by assuming that its exponent is slightly greater than 2 in the amount of 1/6,000,000.[10] But this attempt to modify Newton's theory gave way to the general theory of relativity. In Einstein's theory the orbit of a planet of negligibly small mass *subject solely to the sun's gravitational field* is *not* a perfectly closed ellipse about the sun, as it is in Newton's theory. Instead, the orbit of such a planet is a slowly rotating ellipse.[11] For Mercury, which is the planet closest to the sun, the amount of the perihelion precession thus yielded by general relativity *without* any allowance for the effects of any other planets is about 43 seconds of arc per century. This is in satisfactory agreement with the residual portion of the ob-

[9] Cf. "Mercury, Motion of Perihelion," *Encyclopædia Britannica,* 14th ed., 1929, vol. 15, p. 269.

[10] Pannekoek, *A History of Astronomy,* p. 363.

[11] See A. Einstein, "The Foundations of the General Theory of Relativity," in *The Principle of Relativity, a Collection of Original Memoirs* (New York: Dover Publications, 1952), pp. 163–64; and R. C. Tolman, *Relativity, Thermodynamics and Cosmology* (Oxford: Oxford University Press, 1934), pp. 208–9.

served value that had been left unexplained by the perturbations from other planets in Newton's theory.[12] This agreement prompted the Secretary of the Royal Astronomical Society to write in 1929 that general relativity had "satisfactorily cleared up" the difficulty left unresolved by Newtonian theory.[13] Note that the Newtonian theory deduces a perihelion precession *only* because it takes cognizance of the effects of other planets. By contrast, the relativity theory obtains a *part* of the total precession from the gravitational effect of the sun alone. But the relativistic calculation treated the sun's field as spherically symmetrical. Thereby it neglected any gravitational effects on Mercury arising from such oblateness of the sun as is due to the centrifugal effects of its own rotation.[14]

Suppose that the entire sun rotates with a constant angular velocity equal to the average of that observed on different parts of its surface, i.e., with a period of about 25 days. In that case, centrifugal effects yield an amount of oblateness of the sun's mass distribution that can be neglected with impunity apropos of Mercury's perihelion. But in a 1967 paper, R. H. Dicke and H. M. Goldenberg reported an oblateness of the distribution

[12] See V. L. Ginzburg, "Experimental Verifications of the General Theory of Relativity," in *Recent Developments in General Relativity* (New York: Pergamon Press, 1962), pp. 58–60; and R. H. Dicke, *The Theoretical Significance of Experimental Relativity* (New York: Gordon and Breach, 1964), pp. 27–28.

[13] Cf. "Mercury, Motion of Perihelion," *Encyclopædia Britannica*, p. 269.

[14] H. P. Robertson and T. W. Noonan, *Relativity and Cosmology* (Philadelphia: W. B. Saunders Co., 1968), p. 239; and L. I. Schiff, "Gravitation and Relativity," in D. L. Arm, ed., *Journeys in Science* (Albuquerque: University of New Mexico Press, 1967), pp. 153 and 156.

of the sun's visible brightness corresponding to more than four times the above neglected amount.[15] They interpret this *visual* oblateness as being associated with a corresponding degree of oblateness in the sun's *mass* distribution! And in order to reconcile this amount of solar equatorial bulge with the 25-day period of rotation of the sun's surface, they assume that a core in the sun rotates as rapidly as once in a day or two.

But, alas, according to both Newtonian physics and relativity theory, an equatorial bulge in the sun's mass distribution of the magnitude claimed by Dicke and Goldenberg alone makes for a precession of Mercury's perihelion in the amount of 3.4 seconds of arc per century. (Here the Newtonian calculation is an excellent relativistic approximation!)[16] This amount of rotation of Mercury's ellipse must be added to the relativistic value of 43 seconds, which prevails in the absence of any oblateness. Hence the conjunction of Einstein's equations with the assertion of an oblate mass distribution in the amount assumed by Dicke entails the value of 46.4 seconds of arc per century. Thus, if the sun's mass distribution does deviate from spherical symmetry in the amount claimed by Dicke, and *if*, furthermore, the observations of Mercury's orbit can be taken as proving that the actual value of the unexplained residuum is significantly different from 46.4 seconds, then Einstein's general theory of relativity gives rise to a false conclusion. And, in that case, relativity theory has been refuted in Dicke's laboratory!

[15] R. H. Dicke and H. M. Goldenberg, *Physical Review Letters 9* (1967), p. 313.

[16] See Schiff, "Gravitation and Relativity," p. 156.

It is evident, therefore, that Dicke's disproof of general relativity is predicated, among other things, on the following important assumption: the observed amount of *optical* oblateness of the sun *confirms* it to be *true* that the sun's mass distribution deviates correspondingly from spherical symmetry, i.e., establishes that the sun is also oblate *gravitationally*. To be sure, if gravitational oblateness were present, then it would serve to explain the observed optical flattening. But the existence of optical flattening does *not* entail, and hence cannot guarantee deductively, that there is gravitational oblateness. The inference from optical to gravitational oblateness is *inductive* and uncertain rather than deductive and certain. Indeed, it was pointed out in 1967 that there are at least two possible explanations of the sun's apparent equatorial bulge, neither of which involves any appreciable gravitational oblateness at all. And moreover, it has been argued that Dicke's very high differential rotation rate between the sun's core and surface is too unstable to endure for any length of time.[17]

No wonder, therefore, that there is disagreement in the scientific community as to whether Dicke's findings do disprove Einstein's general relativity. Thus, L. I. Schiff writes:

> At the present time we cannot conclude that the observed solar oblateness invalidates general relativity theory. On the contrary, in view of . . . the difficulties inherent in Dicke and Goldenberg's interpretation of their observations, it seems most reasonable to assume for the present that Einstein's theory is correct.[18]

[17] *Ibid.*, p. 157. See also note 76 at the end of this essay.
[18] *Ibid.*

This episode from the history of recent and current astronomy *suggests* the following quite *fundamental epistemological conclusion*:

> THE IRREMEDIABLE INCONCLUSIVENESS OF THE VERIFICATION OF AN AUXILIARY COMPONENT OF A TOTAL THEORY BY ITS SUPPORTING EVIDENCE IMPOSES A CORRESPONDING LIMITATION ON THE DEDUCIBILITY OF THE CATEGORICAL FALSITY OF THE MAIN COMPONENT OF THE TOTAL THEORY VIA OTHER EVIDENCE ADVERSE TO THE TOTAL THEORY.

In other words, given that the conjunction $H \cdot A$ entails a false prediction, then this fact alone does *not* enable us to deduce the falsity of H itself, unless A is known to be true. But since the truth of A cannot be *guaranteed* by its supporting evidence, the falsity of H is correspondingly uncertain. A two-stage schematization of the logic of our astronomical example will lend greater specificity to this suggested epistemological moral.

I. $\text{GTR} \cdot A$ (which includes DP)

$\downarrow$ (entails)

C (46.4 seconds of arc/century)

But O (43 seconds of arc/century),

so that $O \rightarrow \sim C$

$\therefore \sim (\text{GTR} \cdot A)$

II. $\sim \text{GTR} \vee \sim A$

But also A

$\therefore \sim \text{GTR}$

Let "GTR" be an abbreviation for "general theory of relativity," and let Dicke's postulate of *gravitational* oblateness be called "Dicke's Postulate" or "DP." Furthermore, let us allow for any further premises which may have been used tacitly in the deduction of the conclusion C that Mercury's perihelion precesses by an additional amount of 46.4 seconds of arc per century from the conjunction of GTR with DP. We can do so by combining these further premises with DP into the set *A* of assumptions *auxiliary* to GTR. Finally, let vertical as well as horizontal arrows represent logical entailment, and let "O" be the statement that the actual amount of the *additional* perihelion precession is 43 seconds of arc per century.

Then our schema shows the following: the warrant for the categorical assertibility of the final conclusion that the GTR is false *turns entirely* on the assertibility of the *truth* of A and of the truth of O. For the *entailments* depicted in the schema are all vouchsafed as certain by the principles of deductive logic.

We already saw that DP and hence A rest on an elaborate *inductive* inference from the observations made by Dicke and his colleagues. The uncertainty of that inference issues in an uncertainty regarding the truth of A. Thus, even if the truth of O could be regarded as wholly unproblematic, the categorical falsity of the GTR could here be deduced *only* if A were unquestionably true. But clearly, no one is entitled to claim that A is thus known to be true, and, of course, Dicke would be the last to do so. Moreover, the truth of O is *not* wholly unproblematic any more than that of A. For O is inferred via the elaborate *inductive* reasoning that yields the *total* "observed" precession from which we must first subtract the theoretical amount attributed to the perturbations of other planets, to obtain the residual "observed" amount of about 43 seconds!

But this means that there is even some uncertainty that 46.4 seconds is an incorrect value, i.e., there is some doubt as to the falsity of the conclusion C entailed by the total theory GTR-cum-A!

The acceptance of O and the assertion of Dicke's A as true on the strength of the sun's visual oblateness would issue in the rejection of the GTR as false. But our analysis shows that nonetheless in actual fact the GTR might be true while A is false even though the sun is visually oblate.

More generally, we can see that the inductive acceptance of an actually false auxiliary hypothesis as true may ironically render a purportedly crucial experiment counter-productive.

$T_1\ (H_1 \cdot A_1)$	$T_2\ (H_2 \cdot A_2)$
$\downarrow$	$\downarrow$
C_1	C_2
C_1 is true	C_2 is false

Logical schema of a purportedly crucial experimental disproof of H_2

For suppose that a theory T_1 composed of a major hypothesis H_1 and an auxiliary A_1 entails a consequence C_1, while the conjunction $H_2 \cdot A_2$ constituting a *rival* theory T_2 entails a consequence C_2 incompatible with C_1· . And suppose further that the so-called crucial experiment yields evidence which is taken to be favorable to the truth of C_1 but adverse to the truth of C_2. Those who consider the experiment as crucial will have assumed on inductive grounds that the auxiliaries A_1 and A_2 are each true. But suppose that, in actual fact, A_1 and A_2

are each false, as could readily happen to be the case. Then the presumed experimental falsity of C_2 fully *allows* the following state of affairs: Although it is considered decisively disproven in a crucial experiment, H_2 is actually true. And the conjunction of H_2 with A_2 entailed a false conclusion C_2 *only* because A_2 is false! Moreover, H_1 is actually false, although the conjunction of the false H_1 with the likewise false A_1 does entail the true consequence C_1. Under these circumstances, the *false* H_1 will emerge triumphant in the scientific community over the true H_2. And those who regard the experiment as crucial will mistakenly discard H_2 as *beyond any possible rehabilitation.* In this case, the supposed crucial experiment is surely quite counterproductive. Needless to say, the experiment might alternatively have been recognized as failing to *decide between* H_1 and H_2 if it had yielded a result incompatible with *both* C_1 and C_2.

Thus, the logical situations encountered in our historical examples exhibit the inadequacy of the received textbook account, according to which a component hypothesis can be refuted once and for all. The great French historian of science, Pierre Duhem, pioneered in making us aware of other such episodes in the earlier history of science. And he bequeathed a very influential philosophical legacy to us on the issue of the falsifiability of a scientific hypothesis. In pursuing our central concern with this issue, the remainder of this essay will therefore address itself to Duhem's philosophical legacy. Let us recall the fundamental epistemological moral *suggested* by our example from recent astronomy. Then we can state our central problem more precisely as follows: Can we *ever* justifiably reject a component hypothesis of a theory as irrevocably falsified by observation?

3. Is it NEVER Possible to Falsify a Hypothesis Irrevocably?

In his book *The Aim and Structure of Physical Theory,* Duhem denied the feasibility of crucial experiments in physics. Said he:

> . . . the physicist can never subject an isolated hypothesis to experimental test but only a whole group of hypotheses; when the experiment is in disagreement with his predictions, what he learns is that at least one of the hypotheses constituting this group is unacceptable and ought to be modified; *but the experiment does not designate which one should be changed* (my italics).[19]

Duhem illustrates and elaborates this contention by means of examples from the history of optics. And in each of these cases, he maintains that "If physicists had attached some value to this task,"[20] any one component hypothesis of optical theory such as the corpuscular hypothesis (or so-called emission hypothesis) could have been preserved in the face of seemingly refuting experimental results such as those yielded by Foucault's experiment. According to Duhem, this continued espousal of the component hypothesis could be justified by "shifting the weight of the experimental contradiction to some other proposition of the commonly accepted optics."[21] Here Duhem is maintaining that the

[19] Pierre Duhem, *The Aim and Structure of Physical Theory* (Princeton: Princeton University Press, 1954), p. 187.
[20] *Ibid.*
[21] *Ibid.*, p. 186.

refutation of a component hypothesis H is at least usually no more certain than its verification could be.

In terms of the notation H and A which we have been using, Duhem is telling us that we could blame an experimentally false consequence C of the total optical theory T on the falsity of A while upholding H. In making this claim, Duhem is quite clear that the falsity of *A* no more *follows* from the experimental falsity of C than does the falsity of H. But his point is that this fact does not logically prevent us from postulating that A is false while H is true. And he is telling us that altogether the pertinent empirical facts *allow* us to reject A as false. Hence we are under no deductive logical constraint to infer the falsity of H. In questioning Dicke's purported refutation of the GTR, I gave sanction to this Duhemian contention in this instance precisely on the grounds that Dicke's auxiliary A is *not* known to be true with certainty.

In contemporary philosophy of science, a *generalized* version of Duhem's thesis with its ramifications has been attributed to Duhem and has been highly influential. I shall refer to this elaboration of Duhem's philosophical legacy in present-day philosophy of science as "the D-thesis." But in doing so, my concern is with the philosophical credentials of this legacy, not with whether this attribution to Duhem himself can be uniquely sustained exegetically as against rival interpretations given by Duhem scholars. But I should remark that at least one such scholar, L. Laudan, has cited textual evidence which casts some doubt on this attribution.[22] The pres-

[22] Laurens Laudan, "On the Impossibility of Crucial Falsifying Experiments: Grünbaum on 'The Duhemian Argument'," *Philosophy of Science* 32 (1965): 295–99.

ent philosophical appraisal is intended to supersede some parts of my earlier published critique of the D-thesis. In his Pittsburgh Doctoral Dissertation, Philip Quinn has pointed out that the version of the D-thesis with which I am concerned can be usefully stated in the form of two subtheses D1 and D2, and has argued that Laudan's attributional doubts are warranted only with respect to D2 but not with respect to D1.

The two subtheses are:

D1. No constituent hypothesis H of a wider theory can *ever* be sufficiently isolated from some set or other of auxiliary assumptions so as to be separately falsifiable observationally. H is here understood to be a constituent of a wider theory in the sense that no observational consequence can be deduced from H alone.

It is a corollary of this subthesis that *no* such hypothesis H *ever* lends itself to a crucially falsifying experiment any more than it does to a crucially verifying test.

D2. In order to state the second subthesis D2, we let T be a theory of *any* domain of empirical knowledge, and we let H be *any* of its component subhypotheses, while A is the collection of the remainder of its subhypotheses. Also, we assume that the observationally testable consequence O entailed by the conjunction H · A is taken to be empirically false, because the observed findings are taken to have yielded a result O′ *incompatible* with O. Then D2 asserts the following: For all potential empirical findings O′ of this kind, there exists at least one suitably revised set of auxiliary assumptions A′ such that the conjunction of H with A′ *can be held to be true and explains O′*. Thus D2 claims that H can be held to be true *and* can be used to explain O′ no matter what O′ turns out to be, i.e., *come what may*.

Note that if D2 did not assert that A′ can be held to be true in the face of the evidence no less than H, then H could not be claimed to explain O′ via A′. For premises which are already *known* to be false are not scientifically acceptable as bases for explanation.[23] Hence the part of D2 which asserts that H and A′ can each be held to be *true* presupposes that either they could not be separately falsified or that neither of them has been separately falsified.

In my prior writings on the D-thesis, I made three main claims concerning it:

1. There are quite trivial senses in which D1 and D2 are uninterestingly true and in which no one would wish to contest them.[24]
2. In its non-trivial form, D2 has not been demonstrated.[25]
3. D1 is false, as shown by counterexamples from physical geometry.[26]

[23] For a discussion of the epistemic requirement that explanatory premises must *not* be known to be false, see Ernest Nagel, *The Structure of Science* (New York: Harcourt, Brace and World, 1961), pp. 42–43.

[24] A. Grünbaum, "The Falsifiability of a Component of a Theoretical System," in *Mind, Matter, and Method: Essays in Philosophy and Science in Honor of Herbert Feigl,* P. K. Feyerabend and G. Maxwell, eds. (Minneapolis: University of Minnesota Press, 1966), pp. 276–80.

[25] *Ibid.*, pp. 280–81.

[26] *Ibid.*, pp. 283–95; and A. Grünbaum, *Geometry and Chronometry in Philosophical Perspective* (Minneapolis: University of Minnesota Press, 1968), Chap. III, pp. 341–51. In Chap. III, §9.2, pp. 351–69 of the latter book, I present a counterexample to H. Putnam's *particular* geometrical version of D2. For a brief summary of Putnam's version, see footnote 69.

Since then, Gerald Massey has called my attention to yet another defect of D2 which any proponent of that thesis would presumably endeavor to remedy. Massey has pointed out that, as it stands, D2 attributes *universal* explanatory relevance and power to any one component hypothesis H. For let O* be *any* observationally testable statement *whatever* which is compatible with $\sim$ O, while O′ is the conjunction $\sim O \cdot O^*$. Assume that $(H \cdot A) \rightarrow O$. Then D2 asserts the existence of an auxiliary A′ such that the theoretical conjunction $H \cdot A'$ explains the putative observational finding $\sim O \cdot O^*$. As Joseph Camp has suggested, the proponent of D2 might reply that A′ itself may potentially explain O* *without* H, even though H is essential for explaining $\sim$ O via A′. But the advocate of D2 has no guarantee that he can circumvent the difficulty in this way.

Instead, he might perhaps wish to require that O′ must pertain to *the same kind of phenomena* as O, thereby ruling out "extraneous" findings O*. Yet even if he can articulate such a restriction or provide a viable alternative to it, there is the following further difficulty noted by Massey: D2 gratuitously asserts the existence of a *deductive* explanation for any event whatever. This existential claim is gratuitous. For there *may* be individual occurrences (in the domain of quantum phenomena or elsewhere) which cannot be explained deductively, because of the irreducibly statistical character of its pertinent laws.

Our governing concern here is the question: "Is there *any* component hypothesis H whatever whose falsity we can ascertain?" I shall try to answer this question by giving reasons for now *qualifying* my erstwhile charge that D1 is false. Hence I shall modify the *third* of my earlier contentions about the D-thesis. But before doing so, I must give my reasons for *not* also retracting either of the first two of these contentions in response to the critical literature which they have elicited. These reasons will occupy a number of the pages that follow. The first of my earlier claims was that D1 and D2 are each true in trivial senses which are respectively exemplified by the following two examples, which I had given:[27]

i. Suppose that someone were to assert the presumably false empirical hypothesis H that "Ordinary buttermilk is highly toxic to humans." Then the English sentence expressing this hypothesis could be "saved from refutation" in the face of the observed wholesomeness of ordinary buttermilk by making the following change in the theoretical system constituted by the hypothesis: changing the rules of English usage so that the intension of the term "ordinary buttermilk" is that of the term "arsenic" as customarily understood. In this Pickwickian sense, D1 could be sustained in the case of this particular H.

ii. In an endeavor to justify D2, let someone propose the use of an A′ which is itself of the form

$$\sim H \vee O'.$$

In that case, it is certainly true in standard systems of logic that

$$(H \cdot A') \rightarrow O'.$$

Let us now see by reference to these two examples why I regard them as exemplifications of trivially true versions of D1 and D2 respectively.

[27] Grünbaum, "The Falsifiability of a Component of a Theoretical System," pp. 277–78.

i. *D1*

Here H was the hypothesis that "ordinary buttermilk is highly toxic to humans." When the proponent of D1 was challenged to save this H from refutation, what he did "save" was *not* the proposition H but the *sentence* H expressing it, as reinterpreted in the following respect: Only the term "ordinary buttermilk" was given a new semantical usage, and *no constraint was imposed on its new usage other than that the ensuing reinterpretation turn the sentence H into a true proposition.*

If one does countenance such *unbridled* semantical instability of some of the theoretical language in which H is stated, then one can indeed thereby uphold D1 in the form of Quine's epigram: "Any statement can be held true come what may, if we make drastic enough adjustments elsewhere in the system."[28] But, in that case, D1 turns into a thoroughly unenlightening truism.

I took pains to point out, however, that the commitments of D1 can also be trivially or uninterestingly fulfilled by semantical devices far more sophisticated and restrictive than the unbridled reinterpretation of some of the vocabulary in H. As an example of such a trivial fulfillment of D1 by devices whose feasibility is itself not at all trivial in other respects, I cited the following:

> . . . suppose we had two particular substances I_1 and I_2 which are isomeric with each other. That is to say, these substances are composed of the same elements in the same proportions and with the same molecular weight

[28] W. V. O. Quine, *From a Logical Point of View* (revised ed.), (Cambridge, Mass.: Harvard University Press, 1961), pp. 43 and 41n.

> but the arrangement of the atoms within the molecule is different. Suppose further that I_1 is not at all toxic while I_2 is highly toxic, as in the case of two isomers of trinitrobenzene. [At this point, I gave a footnote citation of 1, 3, 5-trinitrobenzene and of 1, 2, 4-trinitrobenzene respectively.] Then if we were to call I_1 "aposteriorine" and asserted that "aposteriorine is highly toxic," this statement H could also be trivially saved from refutation in the face of the evidence of the wholesomeness of I_1 by the following device: only *partially* changing the meaning of "aposteriorine" so that its intension is the second, highly toxic isomer I_2, thereby leaving the chemical "core meaning" of "aposteriorine" intact. To avoid misunderstanding of my charge of triviality, let me point out precisely what I regard as trivial here. The preservation of H from refutation in the face of the evidence by a *partial* change in the meaning of "aposteriorine" is trivial in the sense of being only a *trivial* fulfillment of *the expectations raised by the D-thesis* [D1]. But, in my view, *the possibility as such of preserving H by this particular kind of change in meaning* is not at all trivial. For this possibility as such reflects a fact about the world: the existence of isomeric substances of radically different degrees of toxicity (allergenicity)![29]

Mindful of this latter kind of example, I emphasized that a construal of D1 which allows itself to be sustained by this kind of alteration of the intension of "trinitrobenzene" is no less trivial *in the context of the expectations raised by the D-thesis* than one which rests its case on calling arsenic "buttermilk."[30] And hence I

[29] Grünbaum, "The Falsifiability of a Component of a Theoretical System," pp. 279–80.
[30] *Ibid.*, p. 279.

was prompted to conclude that "a necessary condition for the nontriviality of Duhem's thesis is that the theoretical language be semantically stable in the relevant respects [i.e., with respect to the vocabulary used in H]."[31]

Mary Hesse took issue with this conclusion in her review of the essay in which I made these claims. There she wrote:

> . . . it is not clear . . . that "semantic stability" *is* always required when a hypothesis is nontrivially saved in face of undermining evidence. The law of conservation of momentum is in a sense saved in relativistic mechanics, and yet the usage of "mass" is changed—it becomes a function of velocity instead of a constant property. But further argument along these lines is idle without more detailed analysis of what it is for a hypothesis to be the "same," and what is involved in "semantic stability."[32]

This criticism calls for several comments.

(a) Hesse calls for a "more detailed analysis of what it is for a hypothesis to be the 'same,' and what is involved in 'semantic stability.' " To this I say that the *primary* onus for providing that more detailed analysis falls on the shoulders of the Duhemian. For he wishes to claim that his thesis D1 is interestingly true. And we saw that if he is to make such a claim, he must surely *not* be satisfied with the mere retention of the *sentence* H in some interpretation or other.

Fortunately, both the proponent and the critic of D1 can avail themselves of an apparatus of distinctions pro-

[31] *Ibid.*, p. 278.

[32] Mary Hesse, *The British Journal for the Philosophy of Science*, 18 (1968): 334.

posed by Peter Achinstein to confer specificity on what it is for the vocabulary of H to remain semantically stable.

Achinstein introduces semantical categories for describing the various possible relationships between properties of X and the term "X." And he appeals to the normal, standard or typical scientific *use* of a term "X" at a given time as distinct from special uses.[33] He writes:

> . . . I must introduce the concept of relevance and speak of a property as relevant for being an X. By this I mean that if an item is known to possess certain properties and lack others, the fact that the item possesses (or lacks) the property in question normally will count, at least to some extent, in favor of (or against) concluding that it is an X; and if it is known to possess or lack sufficiently many properties of certain sorts, the fact that the item possesses or lacks the property in question may justifiably be held to settle whether it is an X.[34] . . .
>
> Two distinctions are now possible. The first is between positive and negative relevance. If the fact that an item has P tends to count more in favor of concluding that it is an X than the fact that it lacks P tends to count against it, P can be said to have more positive than negative relevance for X. The second distinction is between semantical and nonsemantical relevance and is applicable only to certain cases of relevance.
>
> Suppose one is asked to justify the claim that the reddish metallic element of atomic number 29, which is a good conductor and melts at 1083°C, is copper. One reply is that such properties tend to count in and of them-

[33] Peter Achinstein, *Concepts of Science* (Baltimore: Johns Hopkins Press, 1968), pp. 3ff.

[34] *Ibid.*, p. 6.

selves, to some extent, toward classifying something as copper. By this I mean that an item is correctly classifiable as copper solely in virtue of having such properties; they are among the properties which constitute a final court of appeal when considering matters of classification; such properties are, one might say, intrinsically copper-making ones. Suppose, on the other hand, one is asked to justify the claim that the substance constituting about 10^{-4} per cent of the igneous rocks in the earth's crust, that is mined in Michigan, and that was used by the ancient Greeks, is copper. Among the possible replies is *not* that such properties tend to count in and of themselves, to some extent, toward something's being classifiable as copper—that is, it is not true that something is classifiable as copper solely in virtue of having such properties. These properties do not constitute a final court of appeal when considering matters of classification. They are not intrinsically cooper-making ones. Rather, the possession of such properties (among others) counts in favor of classifying something as copper solely because it allows one to infer that the item possesses other properties such as being metallic and having the atomic number 29, properties that are intrinsically copper-making ones, in virtue of which it is classifiable as copper.[35] . . .

Suppose $P^1, \ldots, P_n$ constitutes some set of relevant properties of X. If the properties in this set tend to count in and of themselves, to some extent, toward an item's being classifiable as an X, I shall speak of them as *semantically relevant* for X. If the possession of properties by an item tends to count toward an X-classification solely because it allows one to infer that the item possesses properties of the former sort, I shall speak of such properties as *nonsemantically relevant* for X. This

[35] *Ibid.*, pp. 7–8.

distinction is not meant to apply to all relevant properties of X, for there will be cases on or near the borderline not clearly classifiable in either way.[36]

. . . in the latter part of the eighteenth century, with the systematic chemical nomenclature of Bergman and Lavoisier, the chemical composition of compounds began to be treated as semantically relevant; and . . . chemical composition . . . provided the basis for classification of compounds.

I have used the labels semantical and nonsemantical relevance because X's semantically relevant properties have something to do with the meaning or use of the term "X" in a way that X's nonsemantically relevant properties do not.[37] . . .

Suppose you learn the semantically relevant properties of items denoted by the term "X." Then you will know those properties a possession of which by actual and hypothetical substances in and of itself tends to count in favor of classifying them as ones to which the term "X" is applicable.[38] . . . Properties semantically relevant for X . . . include those that are logically necessary or sufficient. . . .

Consider a term "X" and the properties or conditions semantically relevant for X. It is perfectly possible that there be two different theories in which the term "X" is used, where the same set of semantically relevant properties of X (or conditions for X) are presupposed in each theory (even though other properties attributed to X by these theories, properties not semantically relevant for X, might be different). If so, the term "X" would not mean something different in each theory.[39]

[36] *Ibid.*, pp. 8–9.
[37] *Ibid.*, p. 9.
[38] *Ibid.*, p. 35.
[39] *Ibid.*, p. 101.

Let us now employ Achinstein's distinctions to characterize a semantically stable use of the *sentence* H. I would say that one is engaged in a *semantically stable use* of the term "X," if and only if no changes are made in the membership of the set of properties which are semantically relevant to being an item denoted by the term "X." And similarly for the semantically stable use of the various terms "X_i" ($i = 1, 2, 3 \ldots n$) constituting the vocabulary of the sentence H in which a particular hypothesis (proposition) is expressed, even though these terms will, of course, not be confined to substance words or to the three major types of terms treated by Achinstein.[40] By the same token, if the entire sentence H is used in a semantically stable manner, then the hypothesis H has remained the same in the face of other changes in the total theory. Moreover, employing different terminology, I dealt with the concrete case of geometry and optics in earlier publications to point out the following: when the rejection of a certain presumed law of optics leads to a change in the membership of the properties *non*semantically relevant to a geometrical term "X," this change does *not* itself make for a semantic instability in the use of "X."[41] I shall develop the latter point further below after discussing Hesse's objection.

In the case of my example of the two isomers of trinitrobenzene, it is clear from the very names of the two isomers that the molecular structure is semantically relevant or even logically necessary to being 1, 2, 4-

[40] *Ibid.*, p. 2.

[41] A. Grünbaum, *Philosophical Problems of Space and Time* (New York: Alfred A. Knopf, Inc., 1963), pp. 143–44; and A. Grünbaum, *Geometry and Chronometry . . .*, pp. 314–17.

trinitrobenzene as distinct from 1, 3, 5-trinitrobenzene or conversely. It will be recalled that presumably only the *former* of these two isomers is toxic. The Duhemian should be concerned to be able to uphold the hypothesis "1, 3, 5-trinitrobenzene is toxic" (H) in a semantically stable manner. Hence, if he is to succeed in doing so, his use of the word "1, 3, 5-trinitrobenzene" must leave all of the semantically relevant properties of 1, 3, 5-trinitrobenzene unchanged. And thus it would constitute only a trivial fulfillment of D1 in this case to adopt a *new use* of "1, 3, 5-trinitrobenzene" which suddenly confers *positive* semantic relevance on the particular properties constituting the molecular structure of 1, 2, 4-trinitrobenzene. I do say that upholding H by such a "meaning-switch" is trivial *for the purposes of D1*. But I do *not* thereby affirm at all the biochemical triviality of the fact which makes it possible to uphold H in this particular fashion. For I would be the first to grant that the existence of isomers of trinitrobenzene differing greatly in toxicity is biochemically significant.

(b) Hesse believes that Einstein's replacement of the Newtonian law of conservation of momentum by the relativistic one is a case in which "a hypothesis is non-trivially saved in face of undermining evidence" amid a violation of semantic stability. And she views this case as a counterexample to my claim that semantic stability is a necessary condition for the non-trivial *fulfillment of D1*. The extension of my remarks about the trinitrobenzene example to this case will now serve to show why I do not consider this objection cogent: it rests on a conflation of being non-trivial in some respect or other with being non-trivial vis-à-vis D1.

One postulational base of the special relativistic dynamics of particles combines *formal homologues* of the two principles of the conservation of mass and momentum with the kinematical Lorentz transformations.[42] It is then shown that it is possible to satisfy the two formal conservation principles such that they are Lorentz-covariant, only on the assumption that the mass of a particle depends on its velocity. And the exact form of that velocity-dependence is derived via the requirement that the conservation laws go over into the classical laws for moderate velocities, i.e., that the relativistic mass m assume the value of the Newton mass m_0 for vanishing velocity.[43] This latter fact certainly makes it non-trivial and useful *for mechanics* to use the word "mass" in the case of both Newtonian and relativistic mass. And hence Einstein's *formal* retention of the conservation principles is certainly *not* an instance of an unbridled reinterpretation of them. Yet despite the interesting common feature of the term "mass," and the formal homology of the conservation principles, the two theories disagree here.

Using Achinstein's concept of mere *relevance,* we can say that the Newtonian and relativity theories disagree here as to the membership of the set of properties *relevant* to "mass." For it is clear that in Newton's theory, velocity-*in*dependence is positively *relevant,* in Achinstein's sense, to the term "mass." And it is likewise clear that velocity-*dependence* is similarly positively relevant in special relativistic dynamics. What is unclear is whether velocity *in*dependence and dependence respec-

[42] See, for example, Tolman, *Relativity, Thermodynamics and Cosmology*, pp. 42–45.

[43] See P. G. Bergmann, *Introduction to the Theory of Relativity* (New York: Prentice-Hall, 1946), pp. 86–88.

tively are *semantically* or *non*semantically relevant. Thus, it is unclear whether relativistic dynamics modified the Newtonian conservation principles in semantically relevant ways or preserved them semantically while modifying only the properties *non*semantically relevant to "mass" and "velocity."

Let us determine the bearing of each of these two possibilities on the force of Hesse's purported counterexample to my claim that semantic stability is a necessary condition for the non-triviality of the D-thesis.

In the event of semantic relevance, semantic stability has been violated in a manner already illustrated by my trinitrobenzene example. For in that event, Newton's theory can be said to assert the conservation principles in an interpretation which assigns to the abstract word "mass" the value m_0 as its denotatum, whereas relativity theory rejects these principles as generally false in that interpretation. But a *non*-trivial fulfillment of D1 here would have required the retention of the hypothesis of conservation of momentum in an interpretation which preserves *all* of the properties semantically relevant to "mass," and to "velocity" for that matter. Since this requirement is not met on this construal, the *formal* relativistic retention of this conservation principle cannot qualify as a nontrivial fulfillment of D1. But the success of Einstein's particular semantical reinterpretation is, of course, highly illuminating in other respects.

On the other hand, suppose that relativistic dynamics has modified only the properties which are *non*-semantically relevant to "mass" and "velocity." In that event, its retention of the conservation hypothesis is indeed non-trivial vis-à-vis D1. But on that alternative construal, the relativistic introduction of a velocity-dependence of mass would *not* involve a violation of semantic stability. And then this retention cannot furnish Hesse with a viable counterexample to my claim that semantic stability is necessary for non-triviality vis-à-vis D1.

Finally, suppose that here we are confronted with one of Achinstein's borderline cases such that the distinction between semantic and non-semantic relevance does not apply in the case of mass to the relevant property of velocity-dependence. In that case, we can characterize the transition from the Newtonian to the relativistic momentum conservation law with respect to "mass" as merely a repudiation of the Newtonian claim that velocity-*in*dependence is positively relevant in favor of asserting that velocity-dependence is thus relevant. And then this transition cannot be adduced as a retention of a hypothesis H which violates semantic stability while being non-trivial vis-à-vis D1.

Since there are borderline cases between semantic and non-semantic relevance, it is interesting that Duhem has recently been interpreted as denying that the distinction between them is ever scientifically pertinent. C. Giannoni reads Duhem as maintaining that actual scientific practice always accords the same semantic role to *all* relevant properties.[44] Thus, C. Giannoni writes:

> Duhem is concerned primarily with quantities which are the subject of derivative measurement rather than fundamental measurement. . . . Quantities which are measured by other means than the means originally used in introducing the concept are derivative relative to this

[44] C. Giannoni, "Quine, Grünbaum, and The Duhemian Thesis," *Nous* 1 (1967): 288

> method of measurement. For example, length can be fundamentally measured by using meter sticks, but it also can be measured by sending a light beam along the length and back and measuring the time which it takes to complete the round trip. We can then calculate the length by finding the product of the velocity of light (c) and the time. Such a method of measurement is dependent not only on the measurement of time as is the first type of derivative quantity, but also on the law of nature that the velocity of light is a constant equal to c.[45]
>
> Duhem further substantiates his view . . . by noting that when several methods are available for measuring a certain property, no one method is taken as the absolute criterion relative to which the other methods are derivative in the second sense of derivative measurement noted above [i.e., in Achinstein's sense of being nonsemantically relevant]. Each method is used as a check against the others.[46]

If this thesis of *semantic parity* among *all* of the relevant properties were correct, then one might argue that a semantically stable use of the sentence H would simply not be possible in conjunction with each of two different auxiliary hypotheses A and A′ of the following kind: A and A′ contain *incompatible* law-like statements each of which pertains to one or more entities that are *also* designated by terms in the sentence H. In earlier

[45] *Ibid.*, pp. 286–87.

[46] *Ibid.*, p. 288. The example which Giannoni then goes on to cite from Duhem is one in which a very *weak* electric current was running through a battery and where, therefore, one or more indicators may fail to register its presence. In that case, the current will still be said to flow if one or another indicator yields a positive response. This particular case may well be a borderline one as between semantic and nonsemantic relevance.

publications, I used physical geometry and optics as a test case for these claims. There I considered the actual use of geometrical and optical language in standard relativity physics[47] as well as a hypothetical Duhemian modification of the optics of special relativity.[48] Let me now develop the import of these considerations. We shall see that it runs counter to the thesis of universal semantic *parity* and sustains the feasibility of semantic stability.

The Michelson-Morley experiment involves a comparison of the round-trip times of light along equal closed spatial paths in different directions of an inertial system. As is made explicit in the account of this experiment in standard treatises, the spatial distances along the arms of the interferometer *are measured by rigid rods.*[49] The issue which arose from this experiment was conceived to be the following: Are the round-trip times *along the equal spatial paths* in an inertial frame generally *unequal,* as claimed by the aether theory, or equal as asserted by special relativity? Hence both parties to the dispute agreed that, to within a certain accuracy, equal numerical verdicts furnished by rigid rods were positively semantically relevant to the geometrical relation term "spatially congruent" (as applied to line segments).[50] But the statement of the dis-

[47] Grünbaum, *Geometry and Chronometry . . .*, pp. 314–17.

[48] Grünbaum, *Philosophical Problems of Space and Time,* pp. 143–44.

[49] E.g., in Bergmann, *Introduction to the Theory of Relativity,* pp. 23–26.

[50] Since the aether-theoretically expected time difference in the second order terms is only of the order of 10^{-15} second, allowance had to be made in practice for the absence of a corresponding accuracy in the measurement of the equality of the two arms.

pute shows that the *positive* relevance of the equality of the round-trip times of light to *spatial* congruence was made *contingent* on the particular *law of optics* which would be borne out by the Michelson-Morley experiment. Let X be the relational property of being congruent as obtaining among intervals of space. And let Y be the relational property of being traversed by light in equal round-trip times, as applied to spatial paths as well. Then we can say that the usage of "X" at the turn of the current century was such that the relation Y was only *non*semantically relevant to the relation term "X," whereas identical numerical findings of *rigid rods* —to be called "RR"—were held to be positively semantically relevant to it.

Relativity physics then added the following results pertinent to the kind of relevance which can be claimed for Y: (1) Contrary to the situation in inertial frames, in the socalled "time-orthogonal" *non*-inertial frames the round-trip times of light in different directions along spatially *equal* paths are generally unequal.[51] (2) On a disk rotating uniformly in an inertial system I, there is generally an *inequality* among the transit times required by light departing from the same point to traverse a given closed polygonal path in opposite

---

This is made feasible by the fact that, on the aether theory, the effect of any discrepancy in the lengths of the two arms should *vary*, on account of the earth's motion, as the apparatus is rotated. For details, see Bergmann, *Introduction to the Theory of Relativity*, pp. 24–26, and J. Aharoni, *The Special Theory of Relativity* (Oxford: Oxford University Press, 1959), pp. 270–73. Indeed, slightly unequal arms are needed to produce neat interference fringes.

[51] Grünbaum, *Geometry and Chronometry in Philosophical Perspective*, pp. 316–17.

senses. Thus, suppose that two light signals are *jointly* emitted from a point P on the periphery of a rotating disk and are each made to traverse that periphery in opposite directions so as to return to P. Then the two oppositely directed light pulses are indeed said to traverse *equal spatial distances,* i.e., the relation X is asserted for their paths on the strength of the obtaining of the relation RR. But the one light pulse which travels in the direction *opposite* to that of the disk's rotation with respect to frame I will travel a shorter path in I, and hence will return to the disk point P earlier than the light pulse traveling in the direction in which the disk rotates. Hence the round-trip times of light for these *equal* spatial paths will be *unequal.*[52] (3) Whereas the spatial geometry yielded by a rigid rod on a stationary disk is Euclidean, such a rod yields a hyperbolic non-Euclidean geometry on a rotating disk.[53]

What is the significance of these several results? We see that space intervals which are RR are called "congruent" ("X") in reference frames in which Y is relevant to their being *non*-X, no less than RR intervals are called "X" in those frames in which Y is relevant to their being X! Contrary to Giannoni's general contention, this important case from geometry and optics exhibits the pertinence of the distinction between semantic and non-semantic relevance. Here RR is positively

[52] *Ibid.*, pp. 314–15.

[53] See, for example, C. Møller, *Theory of Relativity* (Oxford: Oxford University Press, 1952), Ch. VIII, §84.

For a rebuttal to John Earman's objections to ascribing a spatial geometry to the rotating disk, see the detailed account of the status of metrics in A. Grünbaum, "Space, Time and Falsifiability," Part I, *Philosophy of Science* 37 (Dec., 1970): 562–65.

semantically relevant to being X. But Y has only *non*-semantical relevance to X, as shown by the fact that while Y is relevant to X in inertial systems, it is relevant to *non*-X in at least two kinds of non-inertial systems. And this refutes the generalization which Giannoni attributes to Duhem, viz., that there is parity of positive *semantic* relevance among properties such as RR and Y, so long as there are instances in which they are all merely positively *relevant* to X.

This lack of *semantic* parity between Y and RR with respect to "X" allows a semantically *stable* geometrical use of "X" on both the stationary and rotating disks, unencumbered by the *incompatibility* of the optical laws which relate the property Y to X in these two different reference frames. Indeed, by furnishing a *tertium comparationis,* this semantic stability confers *physical* interest on the contrast between the incompatible laws of optics in the two disks, and on the contrast between their two incompatible spatial geometries! For in *both* frames, RR is alike centrally semantically relevant to being X, i.e., in each of the two frames, rigid rod coincidences serve alike to determine what space intervals are assigned equal measures $ds = \sqrt{g_{ik}\, dx^i\, dx^k}$. This sameness of RR, which makes for the semantic stability of "X," need not, however, comprise a sameness of the correctional physical laws that enable us to allow computationally for the thermal and other deviations from rigidity.

I used certain formulations of relativity theory as a basis for the preceding account of the properties semantically and nonsemantically relevant to "X" ("spatially congruent"). But this account does not, of course, gainsay the legitimacy of alternative uses of "X" that would issue in alternative, albeit physically equivalent, formulations. For example, the relativistic interpretation of the upshot of the Michelson-Morley experiment could alternatively have been stated as follows: In any inertial system, distances which are equal in the metric based on light-propagation are equal also in the metric which is based on rigid measuring rods.[54] I would say that this particular alternative formulation places Y on a par with RR as a *candidate* for being semantically relevant to "X." Moreover, I have strongly emphasized elsewhere that we are indeed free to formulate certain physical theories alternatively so as to make Y rather than RR semantically relevant to "X."[55]

So much for matters pertaining to trivial fulfillments of D1.

ii. *D2*

It will be recalled that my example of a trivial fulfillment of D2 involved the use of an A′ which is itself of the form $\sim$ H v O′. In that example, the triviality did *not* arise from a violation of semantic stability. Instead, the fulfillment of D2 by the formal validity of the statement $[H \cdot (\sim H \vee O')] \rightarrow O'$ is only trivial because the entailment holds independently of the specific assertive content of H, unless a restrictive kind of entailment is used such as that of the system E of Anderson and Belnap. No matter what H happens to be about, and no matter whether H is substantively relevant to the specific content of O′ or not, O′ will be entailed

[54] This formulation is given in E. T. Whittaker, *From Euclid to Eddington* (London: Cambridge University Press, 1949), p. 63.

[55] Grünbaum, *Geometry and Chronometry . . . , passim.*

by H via the particular A′ used here. Hence H does *not* serve to explain via A′ the facts asserted by O′, thereby satisfying D2 here only trivially, *if at all.*

As Philip Quinn has shown,[56] one of the fallacies which vitiates J. W. Swanson's purported syntactical proof that D2 holds interestingly[57] is the failure to take cognizance of this particular kind of semantic trivialization or even outright falsity of D2.

This brings us to the second of my previously published claims. The latter was that in its non-trivial form, D2 has not been demonstrated by means of D1. Having dealt with several *necessary* conditions for the nontriviality of A′, I therefore went on to comment on a possible *sufficient* condition by writing:

> I am unable to give a formal and completely general *sufficient* condition for the *non*triviality of A′. And, so far as I know, neither the originator nor any of the advocates of the D-thesis [D2] have ever shown any awareness of the need to circumscribe the class of *non*trivial revised auxiliary hypotheses A′ so as to render the D-thesis [D2] interesting. I shall therefore assume that the proponents of the D-thesis intend it to stand or fall on the kind of A′ which we would all recognize as *non-trivial in any given case,* a kind of A′ which I shall symbolize by A'_{nt}.[58]

And I added that D2 was *undemonstrated,* since it does *not* follow from D1 that

$$(\exists A'_{nt})\ [(H \cdot A'_{nt}) \rightarrow O'].$$

[56] Philip Quinn, "The Status of the D-Thesis," Section III, *Philosophy of Science,* vol. 36, No. 4, 1969.

[57] J. W. Swanson, "On the D-Thesis," *Philosophy of Science,* vol. 34 (1967), pp. 59–68.

[58] Grünbaum, "The Falsifiability of a Component of a Theoretical System," p. 278.

Mary Hesse discusses my charge that D2 is a non-sequitur, as well as my erstwhile objections to D1, which I shall qualify below. And she addresses herself to a particular example of mine which was of the following kind: the *original* auxiliary hypothesis A ingredient in the conjunction H · A is highly confirmed quite separately from H, but the conjunction H · A does entail a presumably incorrect observational consequence. Apropos of this example, she writes:

> . . . Grünbaum admits that . . . 'A is only more or less highly confirmed' (p. 289)—a significant retreat from the claim of truth. He thinks nevertheless that it is crucial that confirmation of A can be *separated* from that of H. But surely this is not sufficient for his purpose. For as long as it is not empirically demonstrable, but only likely, that A is true, it will always be possible to reject A in order to save H. . . . Which is where the Duhem thesis came in. It must be concluded that D [the D-thesis] still withstands Grünbaum's assaults. . . .[59]

It is not clear how Hesse wants us to construe her phrase "in order to save H," when she tells us that "it will always be possible to reject A in order to save H." If she intends that phrase to convey merely the same as "with a view to attempting to save H," then her objection pertains only to my critique of D1, which is not now at issue. And in that case, she cannot claim to have shown that the D-thesis "still withstands Grünbaum's assaults" with respect to my charge of non-sequitur against D2. But if she interprets "in order to save H" in the sense of D2 as meaning the same as "with the assurance

[59] Hesse, *The British Journal for the Philosophy of Science* 18: 334–35.

that there exists an A'_{nt} that permits H to explain O′," then I claim that she is fallaciously deducing D2 from the assumed truth of D1. I quite agree that if D1 is assumed to be true *and* IF THERE EXISTS AN A'_{nt} such that H does explain O′ via A'_{nt}, then D1 would guarantee the feasibility of upholding A'_{nt} or H (but not necessarily both) as true. But this fact does not suffice at all to establish that there *exists* such an A'_{nt}! Thus, my charge of non-sequitur against the purported deduction of D2 from D1 does *not* rest on the complaint that the Duhemian cannot assure *a priori* being able to marshal supporting evidence *for* A'_{nt}.

Thus, if Hesse's criticisms were intended, in part, to invalidate my charge of non-sequitur against D2, then I cannot see that they have been successful. And since Philip Quinn has shown[60] that D2, in turn, does *not* entail D1, I maintain that D1 and D2 are logically independent of one another.

This concludes the statement of my reasons for not retracting either of the first two of my three erstwhile claims, which were restated early in this section. Hence we are now ready to consider whether there are any known counterexamples to D1.

Thus, our question is: Is there any component hypothesis H of some theory T or other such that H can be sufficiently isolated so as to lend itself to separate observational falsification? It is understood here that H *itself* does not have any observational consequences. The example which I had adduced to answer this question affirmatively in earlier publications was drawn from physical geometry. I now wish to re-examine my

[60] University of Pittsburgh Doctoral Dissertation (1970).

claim that the example in question is a counterexample to D1. In order to do so, let us consider an arbitrary surface S. And suppose that the Duhemian wants to uphold the following component hypothesis H about S: *if* lengths $ds = \sqrt{g_{ik}\, dx^i\, dx^k}$ are assigned to space intervals by means of *rigid* unit rods, then Euclidean geometry is the metric geometry which prevails physically on the given surface S. Before investigating whether and how this hypothesis H might be falsified, we must be mindful of an important assumption which is ingredient in the antecedent of its if-clause.

That antecedent tells us that the numbers furnished by *rigid* unit rods are to be centrally semantically relevant to the interval measures ds of the theory. Thus, it is a necessary condition for the *consistent* use of rigid rods in assigning lengths to space intervals that any collection of two or more initially coinciding unit solid rods of whatever chemical constitution can thereafter be used *interchangeably* everywhere in the region S, *unless* they are subjected to independently designatable perturbing influences. Thus, the assumption is made here that there is a concordance in the coincidence behavior of solid rods such that no inconsistency would result from the subsequent interchangeable use of initially coinciding unit rods which remain *unperturbed* or "rigid" in the specified sense. In short, there is concordance among rigid rods such that all rigid rods *alike* yield the same metric ds and thereby the same geometry. Einstein stated the relevant empirical law of concordance as follows:

> All practical geometry is based upon a principle which is accessible to experience, and which we will now try

to realise. We will call that which which is enclosed between two boundaries, marked upon a practically-rigid body, a tract. We imagine two practically-rigid bodies, each with a tract marked out on it. These two tracts are said to be "equal to one another" if the boundaries of the one tract can be brought to coincide permanently with the boundaries of the other. We now assume that:

If two tracts are found to be equal once and anywhere, they are equal always and everywhere.

Not only the practical geometry of Euclid, but also its nearest generalisation, the practical geometry of Riemann and therewith the general theory of relativity, rest upon this assumption.[61]

I shall refer to the empirical assumption just formulated by Einstein as "Riemann's concordance assumption," or, briefly, as "R." Note that the assumption R predicates the concordance among rods of different chemical constitution on their being rigid or free from perturbing influences. Perturbing influences are exemplified by inhomogeneities of temperature and by the presence of electric, magnetic and gravitational fields. Thus, two unit rods of different chemical constitution which initially coincide when at the same standard temperature will generally experience a destruction of their initial coincidence if brought to a place at which the temperature is different, and the amount of their thermal elongation will depend on their chemical constitution. By the same token, a wooden rod will shrink or sag more in a gravitational field than a steel one. Since such perturbing forces produce effects of different magnitude on

[61] A. Einstein, "Geometry and Experience," in *Readings in the Philosophy of Science*, H. Feigl and M. Brodbeck, eds. (New York: Appleton-Century-Crofts, 1953), p. 192.

different kinds of solid rods, Reichenbach has called them "differential forces." The perturbing influences which qualify as differential are linked to designatable sources, and their presence is certifiable in ways other than the fact that they issue in the destruction of the initial coincidence of chemically different rods. But the set of perturbing influences is open-ended, since physics cannot be presumed to have discovered *all* such influences in nature by now. This open-endedness of the class of perturbing forces must be borne in mind if one asserts that the measuring rods in a certain region of space are free from perturbing influences or "rigid."

It is clear that if our surface S is indeed free from perturbing influences, then it follows from the assumption R stated by Einstein that *any* two rods of different chemical constitution which initially coincide in S will coincide everywhere else in S, independently of their respective paths of transport. This result has an important bearing on whether it might be possible to falsify the putative Duhemian hypothesis H that the geometry G of our surface S is Euclidean.

For suppose now that S does satisfy the somewhat idealized condition of actually being macroscopically free from perturbing influences. Let us call the auxiliary assumption that S is thus free from perturbing influences "A." Then the conjunction H · A entails that measurements carried out on S should yield the findings required by Euclidean geometry. Among other things, this conjunction entails, for example, that the ratio of the periphery of a circle to its diameter should be π on S. But suppose that the surface S is actually a sphere rather than a Euclidean plane and that the measurements carried out on S with presumedly rigid rods yield various values significantly *less* than π for this ratio, depending on the size of the circle. How then is the Duhemian going to uphold his hypothesis H that if lengths ds on S are measured with rigid rods, the geometry G of S will be Euclidean so that all circles on S will exhibit the ratio π independently of their size?

Clearly, he will endeavor to do so by denying the initial auxiliary hypothesis A, and asserting instead the following: the rods in S are not rigid after all but are being subjected to perturbing influences constituted by differential forces. Obviously, just as in the case of Dicke's refutation of Einstein's theory, the falsification of H itself requires that the truth of A be established, so that Duhem would not be able to deny A. Suppose that we seek to establish A on the strength of the following further experimental findings: However different in chemical appearance, all rods which are found to coincide initially in S preserve their coincidences everywhere in S to within experimental accuracy independently of their paths of transport. How well does this finding establish the truth of A, i.e., that S is *free* from all perturbing influences which would act differentially on rods of diverse chemical constitution?

Let us use the letter "C" to denote the statement that whatever their chemical constitution, any and all rods invariably preserve their initial coincidences under transport in S. Then we can say that Riemann's concordance assumption R states the following: If A is true of *any* region S, then C is true of that region S. Clearly, therefore, C follows from the conjunction A · R, but *not* from A alone. Hence if the observed preservation of coincidence or concordance in S establishes anything here, what it does establish at best is the conjunction

A · R rather than A alone, since I would *not* wish to argue against the Duhemian holist that any finding C which serves inductively to establish a conjunction A · R must be held to establish likewise each of the conjuncts separately.[62] This fact does not, however, give the Duhemian a basis for an objection to my attempt to use the observed concordance in order to establish A. To see this, recall that R must be assumed by the Duhemian if his claim that S is Euclidean is to have the physical significance which is asserted by his hypothesis H. Since the Duhemian wants to uphold H, he could not contest R but only A. Hence, in challenging my attempt to establish A, the Duhemian will *not* be able to object that the observed concordance of the rods in S could establish only the conjunction A · R rather than establishing A *itself*. The issue is therefore one of establishing A. And for the reasons given in Einstein's statement of R above, R is assumed both by the Duhemian, who claims that our surface S is a Euclidean plane, and by the anti-Duhemian, who maintains that S is a sphere.

The Duhemian can argue that the observed concordance cannot conclusively establish A for the following two reasons: (1) He can question whether C can establish A, even if the *apparent* concordance is taken to establish the *universal* statement C indubitably, and (2)

[62] By the same token, if *alternatively* the initial coincidence of the rods had been observed to *cease conspicuously* in the course of transport, the latter observations could *not* be held to falsify A in isolation from R. This inconclusiveness with respect to the falsity of A itself would obtain in the face of the *apparent* discordances, even if the latter are taken as *indubitably* falsifying the claim C of universal physical concordance. For C is entailed by A · R rather than by A alone.

he can challenge the initial inference from the seeming concordance to the physical truth of C. Let me first articulate each of these two Duhemian objections in turn. Thereafter, I shall comment critically on them.

Objection 1

The Duhemian grants that the mere presence of a *single* perturbing influence of significant magnitude would have produced differential effects on chemically different rods. Thus, he admits that the presence of a *single* such influence would be incompatible with C. But he goes on to point out that C *is* compatible with the joint presence of several perturbing influences of possibly unknown physical origin which just *happen* to act as follows: different effects produced on the rods by each one of these perturbing influences are of just the right magnitude to combine into the same *total* deformation of each of the rods. And all being deformed *alike* by the collective action of differential forces *emanating from physical sources,* the rods behave in accord with C. In short, the Duhemian claims that such a *superposition* of differential effects is compatible with C and that it therefore cannot be ruled out by C. Hence he concludes that C cannot be held to establish the truth of A beyond question. Since this Duhemian objection is based on the possibility of conjecturing the specified kind of superposition, I shall refer to it as "the superposition objection."[63]

[63] This kind of objection is raised by L. Sklar, "The Falsifiability of Geometric Theories," *The Journal of Philosophy* 64 (1967): 247–53. Swanson, "On the D-Thesis," presents a model universe for which he conjectures superposition of differential ef-

Objection 2

This second Duhemian objection calls attention to the fact that in inferring C from the *apparent* concordance, we have made the following several inductive inferences: (i) We have taken differences in chemical *appearance* to betoken actual differences in chemical constitution, (ii) we have taken apparent coincidence to betoken actual coincidence, at least to within a certain accuracy, and (iii) we have inductively inferred the *invariable* preservation of coincidence by *all* kinds of rods everywhere in S from only a limited number of trials. The first two of these three inferences invoke collateral hypotheses. For they rest on presumed laws of psychophysics and neurophysiology to the following effect: certain appearances, which are contents of the awareness of the human organism, are lawfully correlated with the objective physical presence of the respective states of affairs which they are held to betoken. Hence let us say that the first two of these three inferences infer a conclusion of the form "Physical item P *has* the property Q" from the premise "P *appears* to be Q," or that they infer *being* Q from *appearing* to be Q.

These objections prompt two corresponding sets of comments.

---

fects *in name only*. For, as Philip Quinn has noted (in "The Status of the D-Thesis"), Swanson effectively adopts the convention that even in the *absence* of perturbing influences, all initially coinciding rods will be held to be *non*-rigid by being assigned lengths that *vary alike* with their positions and/or orientations. Thus, Swanson renders the superposition conjecture physically empty by ignoring the crucial requirement that the alleged differential effects are held to emanate individually from physical sources. Swanson's model universe will be discussed further in footnote 72.

1. It is certainly true that A is *not* entailed by the conjunction R · C. For R is the conditional statement "If A, then C." Since A does not follow deductively from C · R, the premise C · R *cannot* be held to rule out the rival superposition conjecture DEDUCTIVELY. Thus, given R, C *deductively allows* the superposition conjecture as a *conceivable* alternative to A. Nevertheless, we shall now see that in the context of R, the inductive confirmation of A by C is so enormously high that A can be regarded as *well-nigh established* by R · C. Since my impending inductive argument will make use of Bayes' theorem, it will be unconvincing to those who question the applicability of that theorem to the probability of a *hypothesis*. For example, Imre Lakatos registered an objection in this vein.[64]

I shall adopt the non-standard probability notation employed by H. Reichenbach[65] and W. C. Salmon[66] to let "P(L,M)" mean "the probability *from* L to M" or "the probability of M, given L." In the usual notations, the two arguments L and M are reversed.

Recalling that R says "If A, then C" (for any region S), we wish to evaluate the probability P(R·C,A) of A, given that R · C. To do so, we use a special form of Bayes' theorem[67] and write

$$P(R \cdot C,A) = \frac{P(R,A)}{P(R,C)} \times P(R \cdot A,C).$$

[64] See Imre Lakatos and Alan Musgrave, eds., *Criticism and Growth of Knowledge* (New York: Cambridge University Press, 1970), p. 187. But if he countenances entertaining the superposition conjecture ∼A on the grounds that *A* cannot be said to be highly confirmed, how will this ∼A escape being a *metaphysical* proposition?

[65] H. Reichenbach, *The Theory of Probability* (Berkeley: University of California Press, 1949).

[66] W. C. Salmon, *The Foundations of Scientific Inference* (Pittsburgh: University of Pittsburgh Press, 1967), p. 58.

C follows *deductively* from R · A. Hence C is guaranteed by R · A, so that

$$P(R \cdot A, C) = 1.$$

Thus, our formula shows that our desired posterior probability P(R·C,A) will be very high, if the ratio P(R,A)/P(R,C) is itself close to 1, as indeed it will now turn out to be. Needless to say, the *ratio* of these probabilities can be close to 1, even though they are individually quite small.

We can compare the two probabilities in this ratio with respect to magnitude without knowing their individual magnitudes. To effect this comparison, we first consider the conditional probability of C, if we are given that R is true while A is *false*. We can now see that this probability P(R · ~A,C) is very low, whereas we recall that P(R·A,C) = 1. For suppose that A is false, i.e., that S *is* subject to differential forces. Then C can hold only in the *very rare* event that there *happens* to be just the right superposition. Incidentally in the case of our surface S, there is no evidence at all for the existence of the *physical sources* to which the Duhemian needs to attribute his superposed differential effects. Hence we are confronted here with a situation in which the Duhemian wants C to hold under conditions in which superposed differential forces are actually operative, although there is no independent evidence whatever for them.[68]

---

[67] For this form of the theorem, see Reichenbach, *Theory of Probability*, p. 91, equation (6).

[68] This lack of evidence is here construed as obtaining at the given time. An alternative construal in which there avowedly would *never* be any such evidence at all is tantamount to an admission that A is true after all, or that ~A is physically empty. See Philip Quinn ["The Status of the D-Thesis,"] for a discussion of the import of this point for L. Sklar's *particular* superposition objection.

Since $P(R \cdot \sim A,C)$ is very low, we can say that the probability $P(R,C)$ of concordance among the chemically different rods is *only slightly greater, if at all,* than the probability $P(R,A)$ of the occurrence of a region S which is free from differential forces. And this comparison among these probabilities holds here, notwithstanding the fact that there may be as yet undiscovered kinds of differential forces in nature, as I stressed above. It follows that $P(R \cdot C,A)$ is close to 1. And since

$$P(R \cdot C, \sim A) = 1 - P(R \cdot C,A),$$

the conditional probability of $\sim A$ is close to zero.[69]

Thus, the Duhemian's first objection does not detract, and indeed may not be intended to detract from the fact that A is *well-nigh* established by $R \cdot C$.

[69] H. Putnam claims to have shown in "An Examination of Grünbaum's Philosophy of Geometry," in *Philosophy of Science, The Delaware Seminar*, vol. 2, B. Baumrin, ed., (New York: Interscience Publishers, 1963), pp. 247–55, that a superposition of so-called gravitational, electromagnetic, and interactional differential forces can always be invoked here to assert $\sim A$. In particular, he is committed to the following: Suppose A is assumed for a region S (say, on the basis of C) and that a particular metric tensor g_{ik} then results from measurements conducted in S with rods that are presumed to be rigid in virtue of A. Then, says Putnam, we can *always* assert $\sim A$ of S by reference to the specified three kinds of forces. And we can do so in such a way that rods corrected for the effects of these forces would yield any desired different metric tensor g'_{ik}. The latter tensor would enable the Duhemian to assert his chosen geometric H of S.

But I have demonstrated elsewhere in detail [Grünbaum, *Geometry and Chronometry in Philosophical Perspective*, Ch. III, §9] that the so-called gravitational, electromagnetic and interactional forces which Putnam wishes to invoke in order to assert $\sim A$ do *not* qualify *singly* as differential. Hence Putnam's proposed scheme cannot help the Duhemian.

2. The second objection noted that being Q is inferred inductively from appearing to be Q in the case of chemical difference and of coincidence. Furthermore, it called attention to the inductive inference of the *universal* statement C from only a limited number of test cases. What is the force of these remarks?

Let me point out that the claim of mere chemical *difference* does *not* require the definite identification of the particular chemical constitution of each of the rods as, say, iron or wood.[70] Now suppose that the apparent chemically relevant differences among most of the tested rods are striking. And note that they can be further enriched by bringing additional rods to S. Would it then be helpful to the Duhemian to postulate that all of these prima facie differences in chemical appearance mask an underlying chemical identity? If he wishes to avoid resorting to the fantastic superposition conjecture, the Duhemian might be tempted to postulate that the great differences in chemical appearance belie a true crypto-identity of chemical constitution. For by postulating his crypto-identity, he could hope to render the observed concordance among the rods unproblematic for $\sim$A *without* the superposition conjecture. Crypto-identity might make superposition dispensable because the Duhemian might then be able to assert that S is subject to only *one* kind of differential force.[71] But

[70] For some pertinent details, see Grünbaum, "The Falsifiability of a Component of a Theoretical System," pp. 286–87.

[71] It is not obvious that the Duhemian could make this assertion in this context with impunity: even in the presence of only one kind of differential force, there may be rods of one and the same chemical constitution which exhibit the hysteresis behavior of ferromagnetic materials in the sense that their coincidences at a given place will depend on their paths of transport. Cf. Grünbaum, *Geometry and Chronometry . . .*, p. 359.

there is no reason to think that the crypto-identity conjecture is any more probable than the incredible superposition conjecture. And either conjecture may try a working scientist's patience with philosophers.[72]

As for the inductive inferences of physical coincidence from apparent coincidence, the degree of accuracy to which it can be certified inductively is, of course, limited and macroscopic. But what would it avail the Duhemian to assume in lieu of superposition the imperceptibly slight differential deformations which are compatible with the limited accuracy of the observations of coincidence? Surely the imprecision involved here does not provide adequate scope to modify A sufficiently to be able to reconcile the substantial observed deviations

[72] Swanson, "On the D-Thesis," pp. 64–65, maintains that when the Duhemian contests A in my geometric example, he would *not* invoke a crypto-*identity* conjecture, as I have him do, but rather a conjecture of a crypto-*difference* among prima facie chemically *identical* rods. And that crypto-difference might arise from a kind of undiscovered isomerism. He says:

> It is not the assertion of the chemical *difference* of two rods that the D-theorist would claim to be theory-laden. Grünbaum is certainly right in showing that the D-theorist could not argue for *that*. Rather, it is the assertion of their *identity* that would be theory-laden. For suppose that according to a given chemical theory T_1, two rods a and b were found to be chemically identical to one another. . . . We can now construe T_1 as some sort of pre-"isomeric" chemistry . . . and then go on to imagine a post-"isomeric" T_2 such that upon the fulfillment of certain tests . . . it is found that $a \neq b$. . . . But if identity of a and b is thus theory-laden, it does little good to argue that non-identical rods can be clearly discriminated.

Swanson then considers a model universe of three rods, two of which are prima facie chemically identical though crypto-different, while being pairwise objectively and perceptibly different from the third. And he then introduces two (ghostlike) perturbing influences whose *superposition* imparts the same total deformation to each of the three rods. Thus, in the context of his

of the circular ratios from π with the Euclidean H that he needs to save.

What of the inductive inference of the universal statement C from only a limited number of test cases? Note that C need not be established in its full universality in order to create inductive difficulties for the superposition conjecture and to provide strong support for A. For as long as all the chemically different rods actually tested satisfy C, the Duhemian must make his superposition conjecture or crypto-identity assumption credible with respect to *these* rods. Otherwise, he could not reconcile the measurements of circular ratios furnished by them on S with his Euclidean H.

---

superposition model, Swanson attaches significance to the fact that the first two of his three rods, a_1 and a_2, are crypto-different.

But nothing in his argument from superposition depends on the hypothesis that a_1 and a_2 are crypto-different instead of identical, nor does his superposition model gain anything from that added hypothesis. Surely the differential character of his two perturbing influences would not be gainsaid by the mere supposition that a_1 and a_2 are chemically identical and that either of the two perturbing influences therefore affects them alike. If there is superposition issuing in the same total deformation, as contrasted with the absence of differential perturbations, then chemically identical rods can exhibit the same total deformation no less than conspicuously different or crypto-different ones! Far from lending added plausibility to the superposition conjecture, Swanson's introduction of the hypothesis of crypto-difference merely compounds the inductive felonies of the two.

In fact, Swanson overlooked the non-sequitur which I did commit apropos of crypto-identity in my 1966 paper. As is clear from the present essay, the Duhemian could couple his denial of A with the superposition conjecture *rather than* with the hypothesis of crypto-identity. It was therefore incorrect on my part to assert in 1966 that, when denying A, the Duhemian "must" assert crypto-identity (See "The Falsifiability of a Component of a Theoretical System," p. 287, *item (3)*).

The Duhemian might point out that I have inferred A inductively in a two-step inductive inference, which first infers at least instances of the physical claim C from apparent coincidences of prima facie different kinds of rods, and then proceeds to infer A inductively from these instances of C, coupled with R. And he might object that the relation of inductive support is not transitive. It is indeed the case that the relation of inductive support fails to be transitive.[73] But my inductive inference of A does not require an intermediate step via C, for cases of *apparent* coincidence, no less than cases of *physical* coincidence, are correlated with A · R, and R is no more in question here than before. Thus, given R, the inference can proceed to A directly via the assumption that most cases of apparent coincidence of prima facie chemically different rods are correlates of cases of A.[74] And my inductive inference of A is thus predicated on this correlation.

In conclusion, we can now try to answer two questions: (i) To what extent has our verification of A been separate from the assumption of H?, and (ii) to what extent has H *itself* been falsified? When I speak here of a hypothesis as having been "falsified" in asking the latter question, I mean that the *presumption* of its falsity

[73] For a lucid explanation of the *non*-transitivity (as distinct from intransitivity) of the relation of inductive support and of its ramifications, see W. C. Salmon, "Consistency, Transitivity, and Inductive Support," *Ratio* 7 (1965): 164–68.

[74] This claim of direct inductive inferability is made here only with respect to the specifics of this example. And it is certainly not intended to deny the existence of other situations, such as those discussed by R. C. Jeffrey, *The Logic of Decision* (New York: McGraw-Hill, 1965), pp. 155–56, in which a perceptual experience can reasonably prompt us to entertain a variety of relevant propositions.

has been established, *not* that its falsity has been established with certainty or irrevocably. This construal of "falsify" is, of course, implicit in the presupposition of my question that there are different degrees of falsification or differences in the extent to which a hypothesis can be falsified.

(i) Our verification of A did proceed in the context of the assumption of R. And while we saw that R is ingredient in the Euclidean H of our example, as specified, R is similarly ingredient in the rival hypothesis that our surface S is not a Euclidean plane but a spherical surface. Thus, our verification of A was separate from the assumption of the *distinctive* physical content of the particular Duhemian H.

(ii) Duhem attributed the inconclusiveness of the falsification of a component hypothesis H to the legitimacy of denying instead any of the collateral hypotheses A which enter into any test of H. Our analysis has shown that *the denial of A is legitimate precisely to the extent that its* VERIFICATION *suffers from inductive uncertainty.* Moreover, in each of our examples of attempted falsification, the inconclusiveness is attributable *entirely* to the inductive uncertainty besetting the following two *verifications*: the verification of A, and the verification of the so-called observation statement which entails the *falsity* of the conjunction H · A. In short, the inconclusiveness of the falsification of a component H derives wholly from the inconclusiveness of verification. And the falsification of H itself is inconclusive or revocable in the sense that the falsity of H is *not* a *deductive* consequence of premises *all* of which can be known to be true with certainty.

Hence if the falsification of H denied by Duhem's D1

is construed as *irrevocable*, then I agree with Mary Hesse[75] that my geometrical example does not qualify as a counterexample to D1. But I continue to claim that it does so qualify, if one requires only the *very strong presumption* of falsity. And to the extent that my geometrical example does falsify the A'_{nt} which D2 invokes in conjunction with the H of that example, the example also refutes D2's claim that A'_{nt} can justifiably be held to be true.

Subject to an important *caveat* to be issued presently, I maintain, therefore, that there are cases in which we can establish a strong presumption of the falsity of a component hypothesis, although we cannot falsify H in these cases beyond any and all possibility of subsequent rehabilitation. Thus, I emphatically do allow for the following possibility, though not likelihood, in the case of an H which has been falsified in my merely presumptive sense: A daring and innovative scientist who continues to entertain H, albeit as part of a new research program which he envisions as capable of vindicating H, *may* succeed in incorporating H in a theory so subsequently fruitful and well-confirmed as to retroactively alter our assessment of the initial falsification of H. And my *caveat* is that my conception of the falsification of H as establishing the strong presumption of falsity is certainly *not* tantamount to a stultifying injunction to any and all imaginative scientists to cease entertaining H forthwith, whenever such falsification obtains at a given time! Nor is this conception of falsification to be construed as being committed to the historically false assertion that no inductively *un*warranted and daring

[75] Hesse, *The British Journal for the Philosophy of Science* 18:334.

continued espousal of H has ever been crowned with success in the form of subsequent vindication.

Appendix

One might ask: Are there no *other* cases at all in which H can be justifiably rejected as *irrevocably* falsified by observation? On the basis of a schema given by Philip Quinn, it may *seem* that *if* we can ignore such uncertainty as attaches to our observations when testing H, then this question can be answered affirmatively for reasons to be stated presently. When asserting the prima facie existence of this kind of irrevocable falsification as a fact of logic, Quinn recognized that its relevance to actual concrete cases in the pursuit of empirical science is at best very limited. Specifically, Quinn invites consideration of the following kind of logical situation. Let the observation statement O_3 be pairwise *incompatible* with each of the two observation statements O_1 and O_2, so that $O_3 \rightarrow \sim O_1$, and $O_3 \rightarrow \sim O_2$. And suppose further that

$$[(H \cdot A) \rightarrow O_1] \cdot [(H \cdot \sim A) \rightarrow O_2].$$

Assume also that observations made to test H are taken to establish the truth of O_3. Then we can deduce via *modus tollens* that

$$\sim (H \cdot A) \cdot \sim (H \cdot \sim A).$$

Using the law of excluded middle to assert $A \vee \sim A$, we can write

$$(A \vee \sim A) \cdot [\sim (H \cdot A) \cdot \sim (H \cdot \sim A)].$$

The application of the distributive law and the omission

of one of the two conjuncts in the brackets from each of the resulting disjuncts yields

$$[A \cdot \sim (H \cdot A)] \vee [\sim A \cdot \sim (H \cdot \sim A)].$$

Using the principle of the complex constructive dilemma, we can deduce $\sim H \vee \sim H$ and hence $\sim H$.

Although the assumed truth of the observation statement O_3 allows us to deduce the falsity of H in this kind of case, its relevance to empirical science is at best very limited for the following reasons.

In most, if not all, cases of actual science, the conjunction $H \cdot \sim A$ will not be rich enough to yield an observational consequence O_2, if the conjunction $H \cdot A$ does yield an observational consequence O_1. The reason is that the mere denial of A is not likely to be sufficiently specific in content. What, for example, is the observational import of conjoining the *denial* of Darwin's theory of evolution to a hypothesis H concerning the age of the earth? Thus this feature at least severely limits the relevance of this second group of logically possible cases to actual science.

Furthermore, the certainty of the conclusion that H is false in this kind of case rests on certainty that the observational statement O_3 is true. Such certainty is open to question. But *if* we can ignore our doubts on this score, then it would seem that H can be held to have been falsified beyond any possibility of subsequent rehabilitation. Yet, alas, John Winnie has pointed out that despite appearances to the contrary, the H in Quinn's schema does not qualify as a component hypothesis after all.

For Winnie has noted that Quinn's basic premise

$[(H \cdot A) \rightarrow O_1] \cdot [(H \cdot {\sim}A) \rightarrow O_2]$ permits the deduction of

$$H \rightarrow (O_1 \vee O_2).$$

Thus, if $O_1 \vee O_2$ can be held to qualify as an observationally testable statement, then we find that H entails it *without* the aid of the auxiliary A or of ~A. But in that case, Winnie notes that H does not qualify as a constituent of a wider theory in the sense specified in our statement of D1 above.[76]

[76] I am indebted to Philip Quinn and Laurens Laudan for reading a draft of this paper and suggesting some improvements in it. I also wish to thank Wesley Salmon and Allen Janis for some helpful comments and references.

The rapid production of new results in physics requires me to supplement the doubts stated on page 80, concerning the solar oblateness postulated by Dicke. In January, 1971, after this paper had been written, it was pointed out by A. P. Ingersoll and E. A. Spiegel ("Temperature Variation and the Solar Oblateness," *The Astrophysical Journal 163* (1971), pp. 375 and 381) that Dicke and Goldenberg's solar oblateness measurement may be explained alternatively by an equatorial temperature excess of 30°K, located in the high photosphere and low chromosphere, and that presently available data do not permit to distinguish such temperature variation from true oblateness. Furthermore, in the same issue of *The Astrophysical Journal* (January 15, 1971, p. 361), B. Durney has noted that pole–equator differences in temperature, in turn, appear naturally as a result of the interaction of convection with rotation.

An Appreciation

Alvin Thalheimer was born in Baltimore on July 13, 1894. After receiving his A.B. from Harvard, he pursued graduate work in philosophy at Johns Hopkins, receiving his Ph.D. in 1918. His dissertation, *The Meaning of the Terms "Existence" and "Reality,"* was published in 1920. For a time he taught in the Department of Philosophy at Johns Hopkins, but then entered business. In spite of his change of career, and in spite of his many other activities, he continued to occupy himself with philosophic issues and in 1960 published a second book, *Existential Metaphysics.*

However, it was not as philosopher or as businessman that Alvin Thalheimer was best known and will be most gratefully remembered; rather, his influence was most deeply felt through his contribution to a wide variety of civic enterprises and charitable organizations. He was active in the Associated Jewish Charities in Baltimore for a great many years; he was a member of the Baltimore Council of Social Agencies for eight years, and its president for three. In 1957 he was appointed chairman of the Maryland State Welfare Board which he served with distinction until his death. In addition, he was a member of the Visiting Committee for the Department of Philosophy at Harvard University, and for the Harvard Graduate School of Education; he was also a member of the Visiting Committee for the Humanities at Johns Hopkins, and he had a long association with the Baltimore Hebrew College, becoming the chairman of its Board of Trustees in 1963. He died on July 8, 1965.

Fanny Blaustein Thalheimer was born in Baltimore on January 12, 1895, and died on May 28, 1957. She too was active in many educational, civic, and philanthropic organizations, and, like her husband, she too continued to cultivate her early interests, in this case, creative art. After graduating from Western High School in Baltimore she attended the Maryland Institute, and then transferred to Barnard College. In 1918 she and Alvin Thalheimer were married, and she again took up painting at the Maryland Institute, studying with John Sloan and Leon Kroll. In addition to her family responsibilities, Mrs. Thalheimer served for thirteen years, beginning in 1938, on the Maryland State Board of Education; when offered further appointment, she was led by her other obligations to decline. Among these were her service on the Board of Trustees of the Peabody Institute, the Union Memorial Hospital, and the Baltimore Museum of Art, on whose Trustees' Executive Committee she also served. In addition, like her husband, Mrs. Thalheimer devoted much of her effort to the work of charitable organizations, furthering the welfare of the city and the state.

Maurice Mandelbaum

Index